T0184760

Communications
in Computer and Information Science 1328

More information about this series at http://www.springer.com/series/7899

Junhui Li · Andy Way (Eds.)

Machine Translation

16th China Conference, CCMT 2020
Hohhot, China, October 10–12, 2020
Revised Selected Papers

Editors
Junhui Li
Soochow University
Suzhou, China

Andy Way
Dublin City University
Dublin, Ireland

ISSN 1865-0929 ISSN 1865-0937 (electronic)
Communications in Computer and Information Science
ISBN 978-981-33-6161-4 ISBN 978-981-33-6162-1 (eBook)
https://doi.org/10.1007/978-981-33-6162-1

This Springer imprint is published by the registered company Springer Nature Singapore Pte Ltd.
The registered company address is: 152 Beach Road, #21-01/04 Gateway East, Singapore 189721, Singapore

Preface

The China Conference on Machine Translation (CCMT), organized by the Chinese Information Processing Society of China (CIPSC), brings together researchers and practitioners in the area of machine translation, providing a forum for those in academia and industry to exchange and promote the latest development in methodologies, resources, projects, and products, with a special emphasis on the languages in China. CCMT (previously known as CWMT) events have been successfully held in Xiamen (2005, 2011), Beijing (2006, 2008, 2010), Harbin (2007), Nanjing (2009), Xian (2012), Kunming (2013), Macau (2014), Hefei (2015), Urumqi (2016), Dalian (2017), Wuyi (2018), and Nanchang (2019) featuring a variety of activities including an Open Source Systems Development (2006), two Strategic Meetings (2010, 2012), and nine Machine Translation Evaluations (2007, 2008, 2009, 2011, 2013, 2015, 2017, 2018, 2019). These activities have made a substantial impact on advancing the research and development of machine translation in China. The conference has been a highly productive forum for the progress of this area and considered a leading and important academic event in the natural language processing field in China. This year, the 16th CCMT was planned to take place in Hohhot, Inner Mongolia, and was finally moved online due to COVID-19. This conference continued being the most important academic event dedicated to advancing machine translation research. It hosted the 10th Machine Translation Evaluation Campaign, featured two keynote speeches delivered by Graham Neubig (Carnegie Mellon University) and Furu Wei (Microsoft Research Asia), and two tutorials delivered by Hao Zhou (ByteDance), and Long Zhou (Microsoft Research Asia). The conference also organized four panel discussions, bringing attention to the bottleneck of neural machine translation, multimodal machine translation, the frontier of machine translation, and the research and career development for PhD students. A total of 78 submissions (including 34 English papers and 44 Chinese papers) were received for the conference. All the papers were carefully reviewed in a double-blind manner and each paper was evaluated by at least three members of an International Scientific Committee. From the submissions, 13 English papers were accepted. These papers address all aspects of machine translation, including improvement of translation models and systems, translation quality estimation, bilingual lexicon induction, low-resource machine translation, etc. We would like to express our thanks to every person and institution involved in the organization of this conference, especially the members of the Program Committee, the machine translation evaluation campaign, the invited speakers, the local organization team, our generous sponsors, and the organizations that supported and promoted the event. Last but not least, we greatly appreciate Springer for publishing the proceedings.

October 2020   Junhui Li

Organization

General Chair

Ming Zhou   Microsoft Research Asia, China

Program Committee Co-chairs

Junhui Li   Soochow University, China
Andy Way   Dublin City University, Ireland

Evaluation Chair

Muyun Yang   Harbin Institute of Technology, China

Organization Co-chairs

Guanglai Gao   Inner Mongolia University, China
Hongxu Hou   Inner Mongolia University, China

Tutorial Co-chairs

Shujie Liu   Microsoft Research Asia, China
Tong Xiao   Northeastern University, China

Student Forum Co-chairs

Fei Huang   Alibaba, China
Shujian Huang   Nanjing University, China

Front-Trends Forum Co-chairs

Yang Liu   Tsinghua University, China
Jiajun Zhang   Institute of Automation, Chinese Academy of Sciences, China

Workshop Co-chairs

Zhaopeng Tu   Tencent, China
Jinsong Su   Xiamen University, China

Publication Chair

Yating Yang   Xinjiang Technical Institute of Physics and Chemistry,
                            Chinese Academy of Sciences, China

Sponsorship Co-chairs

Changliang Li   Kingsoft, China
Yang Feng   Institute of Computing Technology, Chinese Academy
                            of Sciences, China

Publicity Co-chairs

Chong Feng   Beijing Institute of Technology, China
Zhongjun He   Baidu, Inc., China

Program Committee

Hailong Cao   Harbin Institute of Technology, China
Kehai Chen   NICT, Japan
Boxing Chen   Alibaba, China
Jiajun Chen   Nanjing University, China
Yong Cheng   Google, Inc., USA
Jinhua Du   Dublin City University, Ireland
Xiangyu Duan   Soochow University, China
Yang Feng   Institute of Computing Technology, Chinese Academy
                            of Sciences, China
Shengxiang Gao   Kunming University of Science and Technology, China
Zhengxian Gong   Soochow University, China
Yanqing He   Institute of Scientific and Technical Information
                            of China, China
Zhongjun He   Baidu, Inc., China
Guoping Huang   Tencent AI Lab, China
Shujian Huang   Nanjing University, China
Yves Lepage   Waseda University, Japan
Junhui Li   Soochow University, China
Maoxi Li   Jiangxi Normal University, China
Liangyou Li   Huawei Noah's Ark Lab, China
Yaochao Li   Soochow University, China
Qun Liu   Huawei Noah's Ark Lab, China
Yang Liu   Tsinghua University, China
Cunli Mao   Kunming University of Science and Technology, China
Haitao Mi   Ant Financial US, USA
Toshiaki Nakazawa   The University of Tokyo, Japan
Xing Shi   University of Southern California, USA
Kai Song   Alibaba, China

Jinsong Su	Xiamen University, China
Zhaopeng Tu	Tencent AI Lab, China
Mingxuan Wang	ByteDance, China
Rui Wang	NICT, Japan
Shaonan Wang	National Laboratory of Pattern Recognition, Institute of Automation, Chinese Academy of Sciences, China
Xing Wang	Tencent AI Lab, China
Derek F. Wong	University of Macau, Macau, China
Tong Xiao	Northeastern University, China
Hongfei Xu	Saarland University, Germany
Muyun Yang	Harbin Institute of Technology, China
Yating Yang	Xinjiang Technical Institute of Physics and Chemistry, Chinese Academy of Sciences, China
Heng Yu	Alibaba, China
Dakun Zhang	SYSTRAN, France
Jiajun Zhang	Institute of Automation, Chinese Academy of Sciences, China
Xiaojun Zhang	University of Stirling, UK
Tiejun Zhao	Harbin Institute of Technology, China
Renjie Zheng	Oregon State University, USA
Muhua Zhu	Tencent News, China

Organizer

Chinese Information Processing Society of China, China

Co-organizer

Inner Mongolia University, China

Sponsors

Diamond Sponsors

Kingsoft AI

NiuTrans Research

Global Tone Communication Technology Co., Ltd

Gold Sponsors

Baidu

Youdao

Contents

Transfer Learning for Chinese-Lao Neural Machine Translation with Linguistic Similarity

Zhiqiang Yu[1,2], Zhengtao Yu[1,2(✉)], Yuxin Huang[1,2], Junjun Guo[1,2], Zhenhan Wang[1,2], and Zhibo Man[1,2]

[1] Faculty of Information Engineering and Automation,
Kunming University of Science and Technology, Kunming, China
ztyu@hotmail.com
[2] Yunnan Key Laboratory of Artifcial Intelligence,
Kunming University of Science and Technology, Kunming, China

Abstract. As a typical low-resource language pair, besides severely limited by the scale of parallel corpus, Chinese-Lao language pair also has considerable linguistic differences, resulting in poor performance of Chinese-Lao neural machine translation (NMT) task. However, compared with the Chinese-Lao language pair, there are considerable cross-lingual similarities between Thai-Lao languages. According to these features, we propose a novel NMT approach. We first train Chinese-Thai and Thai-Lao NMT models wherein Thai is treated as pivot language. Then the transfer learning strategy is used to extract the encoder and decoder respectively from the two trained models. Finally, the encoder and decoder are combined into a new model and then fine-tuned based on a small-scale Chinese-Lao parallel corpus. We argue that the pivot language Thai can deliver more information to Lao decoder via linguistic similarity and help improve the translation quality of the proposed transfer-based approach. Experimental results demonstrate that our approach achieves 9.12 BLEU on Chinese-Lao translation task using a small parallel corpus, compared to the 7.37 BLEU of state-of-the-art Transformer baseline system using back-translation.

Keywords: Transfer learning · Chinese-Lao · Neural machine translation · Linguistic similarity

1 Introduction

Chinese-Lao NMT is a typical low-resource NMT, the research on which in the past decade is not widespread. Limited by the scale and domain of parallel corpus, the bulk of research on Chinese-Lao NMT has to focus on language model training and dictionary building [1, 2], etc. However, with the introduction of the "the Belt and Road", the demand for translations of Chinese-Lao has been increasing. Therefore, it is important to investigate how to design an effective NMT model on a small scale of parallel corpus to improve translation performance on Chinese-Lao language pair.

To tackle the inefficiency problem in low-resource settings such as Chinese-Lao, some approaches have been proposed. Recent efforts [3–6] in NMT research have shown promising results when transfer learning techniques are applied to leverage

© Springer Nature Singapore Pte Ltd. 2020
J. Li and A. Way (Eds.): CCMT 2020, CCIS 1328, pp. 1–10, 2020.
https://doi.org/10.1007/978-981-33-6162-1_1

existing rich-resource models to cope with the scarcity of training data in low-resource settings. However, these works mainly leverage the way that transfers the parameters of the rich-resource model to the low-resource model, barely adopt the strategy to extract the encoder or decoder from two pivot-relevant models separately. Even when the pivot strategy [6] is adopted, the similarity between the pivot and the target language is ignored.

Chinese and Lao have a mass of linguistic differences, the former belongs to Sino-Tibetan language family and the latter is from the Tai-Kadai language family. The tremendous cross-lingual different make Chinese and Lao are mutually unintelligible. Therefore, we choose a pivot language to overcome such cross-lingual different. Intuitively, a good pivot language for Chinese-Lao translation should have the following properties: (1) Adapt to unbalanced data set: the scale of Chinese-pivot parallel corpus could be larger than Chinese-Lao; and (2) considerable similarities with Lao: has high cross-lingual similarities, the best is in the same language family with Lao. Based on above considerations, we choose Thai as the pivot language for our transfer learning approach, and we elaborate the Language features in next section. Our main contributions are as follows:

- we investigate the cross-lingual similarities between Thai and Lao, and discuss the feasibility that chooses Thai as the pivot language for Chinese-Lao translation model construction.
- we propose a transfer Learning approach for Chinese-Lao NMT with pivot language. The central idea is to construct a new model by extracting encoder from the trained Chinese-Thai NMT model, and decoder from Thai-Lao NMT model which is trained on small scale parallel corpus of high similarity.

2   Linguistic Similarity Between Thai and Lao

Thai and Lao are tonal languages and belong to Tai-Kadai language family, the speech and writing of the two languages are highly similar. Actually, spoken Thai and Lao are mutually intelligible. Moreover, the two languages share a large amount of correlative words on etymologically and have similar head-initial syntactic structures [8]. For writing, Thai and Lao are both written in abugida tokens, and in many cases the sentences composed of which are linguistically similar [9]. As the example illustrated in Fig. 1, the similarity in the shape of certain tokens can be observed.

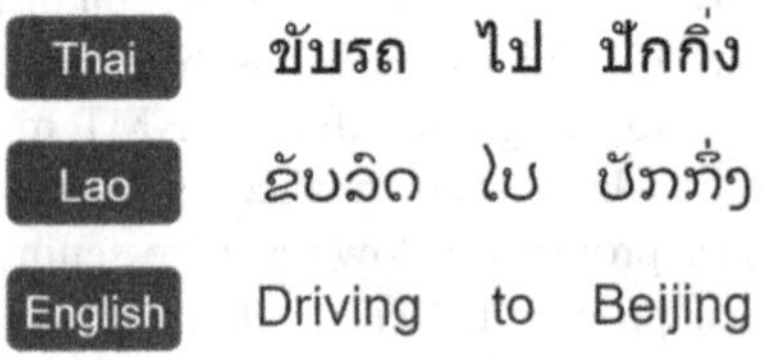

Fig. 1. Thai-Lao linguistic similarity

Besides the similarity investigation of token shape, we also discuss the similarity of syntactic structure. We leverage GIZA++ tool [10] to get word alignment over the 20 K Thai-Lao portion of publicly ALT dataset. Then we use the approach proposed in Isozaki [11] to get the Kendall's τ according to the previous word alignment. Kendall's τ mainly indicate the cost of adjusting two parallel sentences to the same word order. As shown in Fig. 2, Thai-Lao language pair shows a relative similar order with an average τ around 0.73. The result demonstrates the considerable similarity in syntactic structure of the two language.

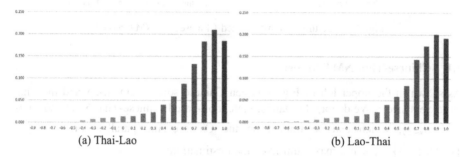

(a) Thai-Lao   (b) Lao-Thai

Fig. 2. Distribution of Kendall's τ on Thai-to-Lao (a) and Lao-to-Thai (b).

According to the above analysis, Thai-Lao language pair has considerable cross-lingual similarity in either token shape or syntactic structure. We argue that the similarity between two languages will bring more adequate information from Thai to Lao and improve the accuracy of Lao decoder. Therefore, choose Thai as the pivot language for Chinese-Lao translation task is positive. To the best of our knowledge, there is no existing work on transfer learning for Chinese-Lao NMT by choosing a target-similar pivot language Thai.

3 Our Approach

In this section, we will elaborate the detail of our proposed model. Our goal is to achieve a transfer-based NMT model which composed of trained Chinese encoder and Lao decoder. As illustrated in Fig. 3, we first train Chinese-Thai and Thai-Lao translation model respectively. Then we compose new translation model using extracted Chinese encoder and Lao decoder. Lastly, we fine-tune the new Chinese-Lao model on small parallel corpus.

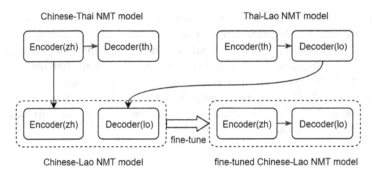

Fig. 3. Structure of transfer-based Chinese-Lao NMT model.

3.1   Chinese-Thai NMT Model

As shown in the upper left of Fig. 3. Given Chinese source sentence x and the Thai target sentence z. We denote the standard attention-based Chinese-Thai NMT model as $P(z|x; \theta_{x \to z})$, which can be trained on the Chinese-Thai parallel corpus $D_{x,z} = \left\{ \langle x^{(m)}, z^{(m)} \rangle \right\}_{m=1}^{M}$ using maximum likelihood estimation:

$$\hat{\theta}_{x \to z} = \underset{\theta_{x \to z}}{\arg \max} \{ \mathcal{L}(\theta_{x \to z}) \} \tag{1}$$

where the log-likelihood is defined as:

$$\mathcal{L}(\theta_{x \to z}) = \sum_{m=1}^{M} \log P\left(z^{(m)} | x^{(m)}; \theta_{x \to z} \right) \tag{2}$$

3.2   Thai-Lao NMT Model

The standard attention-based Thai-Lao NMT model $P(y|z; \theta_{z \to y})$ with respect to the Thai-Lao parallel corpus $D_{z,y} = \left\{ \langle z^{(n)}, y^{(n)} \rangle \right\}_{n=1}^{N}$ can be calculated similar with the Chinese-Thai NMT model, the model training procedure using maximum likelihood estimation is:

$$\hat{\theta}_{z \to y} = \underset{\theta_{z \to y}}{\arg \max} \{ \mathcal{L}(\theta_{z \to y}) \} \tag{3}$$

where the log-likelihood is defined as:

$$\mathcal{L}(\theta_{z \to y}) = \sum_{n=1}^{N} \log P\left(y^{(n)} | z^{(n)}; \theta_{z \to y} \right) \tag{4}$$

3.3   Chinese-Lao NMT Model

We compose a new translation model using the Chinese encoder $Enc_{x \to z}$ and the Lao decoder $Dec_{z \to y}$ that fetched from Chinese-Thai NMT model $P(z|x; \theta_{x \to z})$ and Thai-Lao NMT model $P(y|z; \theta_{z \to y})$ respectively. The process can be simply formulated as:

$$enc_{x \to z} = fetchEnc(P(z|x; \theta_{x \to z})) \tag{5}$$

$$dec_{z \to y} = fetchDec(P(y|z; \theta_{z \to y})) \tag{6}$$

$$P(y|x; \theta_{x \to y}) = \{enc_{x \to z}, dec_{z \to y}\} \tag{7}$$

where *fetchEnc* and *fetchDec* are the functions that fetch encoder and decoder portion parameters from Chinese-Thai and Thai-Lao NMT model respectively. $P(y|x; \theta_{x \to y})$ is the composed Chinese-Lao NMT model.

Even in low-resource NMT settings, there often exist small-scale parallel corpus. In our approach, we first combine the extracted encoder and decoder into a new NMT model, but the model is not fine-tuned and suboptimal. Therefore, we use the small-scale parallel corpus of Chinese-Lao from ALT dataset to fine-tune the model parameters. In the fine-tuning process, we first try to fix some parameters. However, we observe that for the ALT dataset we used for fine-tuning, not fix parameters had a better effect.

4   Evaluation

4.1   Experimental Setup

Data. We conduct experiments on the publicly ALT dataset[1] and the in-house Chinese-Thai parallel corpus. For ALT dataset, we use the trilingual Chinese-Thai-Lao portion which comprise 20K sentences triples. Then we bin the ALT subset into three subsets: 19K for training, other two subsets of 500 sentences as the development and test datasets, respectively. For Chinese-Thai model training, we use the combined parallel data from the Chinese-Thai portion (19K) of ALT subset and the 50K in-house parallel corpus collected by ourselves. For Thai-Lao model training, we use the 19K parallel data from the Thai-Lao portion of ALT subset. For Chinese-Lao model fine-tuning training, we use the 19K parallel data from the Chinese-Lao portion of the ALT subset. We process the experiment corpus simply before applying our approach. For the Thai word segmentation, we use pythaipiece tool[2] which based on sentencepiece to segment Thai sentences, while for Lao word segmentation, we use LaoWordSegmentation tool[3]

---

[1] http://www2.nict.go.jp/astrec-att/member/mutiyama/ALT/.

[2] https://github.com/wannaphong/thai-word-segmentation-sentencepiece.

[3] https://github.com/djkhz/LaoWordSegmentation.

to segment Lao sentences. For Chinese we apply word segmentation by jieba tools[4]. We do not use BPE approach on the experimental parallel corpus.

Evaluation. We adopt the case insensitive 4-gram BLEU as the main evaluation metrics [12], and choose the multi-bleu.perl as scoring script. Significance tests are conducted based on the best BLEU results by using bootstrap resampling [13].

Baseline. We compare the proposed model against the state-of-the-art NMT system Transformer, which has obtained the state-of-the-art performance on machine translation and predicts target sentence from left to right relying on self-attention [14].

Implement Detail. We adopt the prudent Transformer settings, uses a 2-layer encoder and 2-layer decoder, while each layer employs 4 parallel attention heads. The dimensions of word embeddings, hidden states and the filter sizes are set to 256, 256 and 512 respectively. The dropout is 0.2 for Chinese-Thai and 0.1 for Thai-Lao training. We train using the Adam optimizer [15] with a batch size of 256 words and evaluate the model every 1000 steps. The models are trained on 2 P100 GPUs. We implement our approach on Thumt [16], an efficient open source machine translation platform.

4.2 Experimental Results

Quantitative Study. Table 1 shows the experimental results evaluated by BLUE score. We get 3.62 BLEU point improvement compared with transformer baseline which only use 19 K tiny Chinese-Lao parallel corpus for training. Moreover, for a fair comparison, we back-translation [17] the Thai side sentences of 50 K Chinese-Thai corpus collected by ourselves to corresponding Lao sentences on the Thai-Lao transformer model which is trained on original 19 K ALT corpus. The Chinese sentences and corresponding translated Lao sentences are combined as new parallel corpus, which is fed into a new Chinese-Lao model together with Chinese-Lao ALT dataset for training again. Note that we do not back-translation the Lao side sentences of Thai-Lao ALT corpus to Chinese because ALT corpus is a Multilingual parallel corpus. As shown in Table 1, our approach still gains 1.75 BLEU point improvement compared with Transformer using back-translation.

Table 1. BLEU scores evaluated on test set (0.5 K) compared with baseline. Parallel sentences for Transformer and transformer + back-translation training are 19 K and 69 K, respectively.

Models	BLEU
Transformer	5.50
Transformer + back-translation	7.37
Our approach	9.12

---

[4] https://github.com/fxsjy/jieba.

Table 2. Performance difference on our proposed approach when choosing different pivot language.

Pivot	BLEU
En	7.55
Th	9.12

The source side Chinese sentences for Chinese-Thai model training and Chinese-Lao model fine-tuning are identical. To dispel the concern that the improvement is brought by the same source training data, for a fair comparison, we also conduct the experiment that select different language as pivot. To ensure fair comparison, we choose English as the pivot, and for Chinese-English model training, we conduct the experiment on the trilingual Chinese-English-Lao portion of ALT dataset and the extracted 50 K Chinese-English parallel sentences from IWSLT15 zh-en dataset. Table 2 reports the performance of choosing English and Thai as pivot language respectively. We observe that there is large gap when choosing English as pivot compared with Thai. The main possible reason is that English has few cross-lingual similarities with Lao compared with Thai.

Case Study. Apart from the quantitative analysis, we illustrate an example of our propose approach. As we do not apply BPE to corpus, to avoid the UNK, we provide a relatively common sentence. As shown in Table 3, the Chinese word "那人" (*The man*), "处理" (*deal with*) and "秘密的" (*secret*) are translated correctly in our approach. We argue that one of the main reasons is that the pivot language Thai delivers more information in translation process. As shown in Table 4, for the three preceding Chinese words, the corresponding words in Lao are similar in morphology with the words in the pivot language Thai and all of them can be found in the training corpus.

Table 3. Example of Chinese-Lao translation.

Input:	那人 能 处理 秘密的 私人 问题
Golden:	ຜູ້ຊາຍ ສາມາດ ຈັດການກັບ ຄວາມລັບ ບັນຫາສ່ວນຕົວ
	(The man can deal with secret personal problems)
Baseline:	ຜູ້ຊາຍ ສາມາດ ໄດ້ຮັບ ບາງ ບັນຫາ
	(The man often gets some trouble)
Our approach:	ຜູ້ຊາຍ ສາມາດ ຈັດການກັບ ຄວາມລັບ ສິ່ງຂອງ
	(The man often deals with secret things)

Table 4. The morphological similarity between Thai and Lao words that translated correctly in Table 3.

Thai	Lao	Corresponding English
ผู้ชาย	ຜູ້ຊາຍ	the man
จัดการกับ	ຈັດການກັບ	deal with
ความลับ	ຄວາມລັບ	secret

5   Related Work

Many types of transfer learning approaches [3–7] have been proposed in the past few years. Since the advent of Transformer, To improve the quality of the translation, many authors have endeavored to adopt transfer-based method on Transformer framework. Lakew et al. propose a Transformer-adapted transfer learning approach [5] that extend an initial model for a given language pair to cover new languages by adapting its vocabulary as long as new data become available. Kim et al. propose three methods to increase the relation among source, pivot, and target languages in the pre-training and implement the models on Transformer [7]. While for Chinese-Lao translation task, limited by the scale of parallel corpus and the language processing tools of Lao, the research on Chinese-Lao NMT in the past decade is not widespread. The bulk of researches have to focus on the Analysis of Lao Language characteristics [1, 2]. Different from the above work, we endeavor to leverage the cross-lingual similarity between Thai and Lao to improve Chinese-Lao NMT performance based on Transformer framework.

6   Conclusions

We propose a new NMT approach focusing on language pair Chinese-Lao with an extremely limited amount of parallel corpus. Our proposed approach utilizes a transfer learning approach to reuse the encoder and decoder from two trained Chinese-Thai and Thai-Lao NMT models respectively. As the pivot language, Thai has considerable similarities with Lao, and we argue that it will bring significant improvement to entire framework. We conduct contrast experiments, as the results reported, our approach can achieve 9.12 BLEU on Chinese-Lao translation task using small parallel corpus, compared to the 7.37 BLEU of strong transformer baseline system using back-translation.

An interesting direction is to apply our approach to other low-resource NMT task, with the feature that the scale of source-pivot parallel corpus is obvious larger than pivot-target parallel corpus, and the pivot language is similar with target language, such as Chinese-Indonesian (Malay as pivot language) etc.

Acknowledgements. We would like to thank the anonymous reviewers for their constructive comments. The work is supported by National Natural Science Foundation of China (Grant Nos. 61732005, 61672271, 61761026, 61762056 and 61866020), National key research and development plan project (Grant No. 2019QY1800), Yunnan high-tech industry development project (Grant No. 201606), and Natural Science Foundation of Yunnan Province (Grant No. 2018FB104).

References

1. Srithirath, A., Seresangtakul, P.: A hybrid approach to lao word segmentation using longest syllable level matching with named entities recognition. In: The 10th International Conference on Electrical Engineering/Electronics, Computer, Telecommunications and Information Technology (ECTI-CON), Krabi, Thailand, pp. 1–5, 15–17 May 2013
2. Yang, M., Zhou, L., Yu, Z., Gao, S., Guo, J.: Lao named entity recognition based on conditional random fields with simple heuristic information. In: The 12th International Conference on Fuzzy Systems and Knowledge Discovery (FSKD), Zhangjiajie, pp. 1426–1431 (2015)
3. Zoph, B., Yuret, D., May, J., et al.: Transfer learning for low-resource neural machine translation. arXiv preprint arXiv:1604.02201 (2016)
4. Nguyen, T.Q., Chiang, D.: Transfer learning across low-resource, related languages for neural machine translation. arXiv preprint arXiv:1708.09803 (2017)
5. Lakew, S.M., Erofeeva, A., Negri, M., et al.: Transfer learning in multilingual neural machine translation with dynamic vocabulary. arXiv preprint arXiv:1811.01137 ACM Trans. Graph (2018)
6. Cheng, Y., Yang, Q., Liu, Y., et al.: Joint training for pivot-based neural machine translation. In: 26th International Joint Conference on Artificial Intelligence, Morgan Kaufmann, San Francisco, pp. 3974–3980 (2017)
7. Kim, Y., et al.: Pivot-Based Transfer Learning for Neural Machine Translation Between Non-English Languages. In: Proceedings of the 2019 Conference on Empirical Methods in Natural Language Processing and the 9th International Joint Conference on Natural Language Processing (EMNLP-IJCNLP) (2019). n. pag. Crossref. Web
8. Ding, C., Utiyama, M., Sumita, E.: Similar southeast asian languages: corpus-based case study on thai-laotian and Malay-Indonesian. In: Proceedings of the 3rd Workshop on Asian Translation, pp. 149–156, Osaka, Japan, 11–17 December 2016
9. Singvongsa, K., Seresangtakul, P.: Lao-Thai machine translation using statistical model. In: International Joint Conference on Computer Science & Software Engineering. IEEE (2016)
10. Och, F.J., Ney, H.: A systematic comparison of various statistical alignment models. Comput. Linguist. **29**(1), 19–51 (2003)
11. Isozaki, H., Sudoh, K., Tsukada, H., Duh, K.: HPSG-based preprocessing for English-to-Japanese translation. ACM Trans. Asian Lang. Inf. Process. **11**(3), 1–16 (2012)
12. Papineni, K., Roukos, S., Ward, T., Zhu, W.: BLEU: a method for automatic evaluation of machine translation. In: Proceedings of ACL, pp. 311–318 (2002)
13. Koehn, P.: Statistical significance tests for machine translation evaluation. Proc. EMNLP **2004**, 388–395 (2004)
14. Vaswani, A., et al.: Attention is all you need. In: Guyon, I., (ed.) Advances in Neural Information Processing Systems, vol. 30, pp. 5998–6008 (2017)

15. Kingma, D.P., Ba, J.: Adam a method for stochastic optimization. In: Proceedings of the 3rd International Conference on Learning Representations (2015)
16. Zhang, J., et al.: THUMT: An Open Source Toolkit for Neural Machine Translation (2017). arXiv:1706.06415
17. Sennrich, R., Haddow, B., Birch, A.: Improving Neural Machine Translation Models with Monolingual Data (2015)

MTNER: A Corpus for Mongolian Tourism Named Entity Recognition

Xiao Cheng, Weihua Wang$^{(\boxtimes)}$, Feilong Bao, and Guanglai Gao

College of Computer Science, Inner Mongolia University,
Inner Mongolia Key Laboratory of Mongolian Information Processing
Technology, Huhhot 010021, China
yychengxiao99@163.com, wangwh@imu.edu.cn

Abstract. Name Entity Recognition is the essential tool for machine translation. Traditional Named Entity Recognition focuses on the person, location and organization names. However, there is still a lack of data to identify travel-related named entities, especially in Mongolian. In this paper, we introduce a newly corpus for Mongolian Tourism Named Entity Recognition (MTNER), consisting of 16,000 sentences annotated with 18 entity types. We trained in-domain BERT representations with the 10 GB of unannotated Mongolian corpus, and trained a NER model based on the BERT tagging model with the newly corpus. Which achieves an overall 82.09 F1 score on Mongolian Tourism Named Entity Recognition and lead to an absolute increase of +3.54 F1 score over the traditional CRF Named Entity Recognition method.

Keywords: Named entity recognition · Mongolian tourism corpus · NER model based on BERT

1 Introduction

Recently there has been significant interest in modeling human language together with the special domain, especially tourism, as more data become available on websites such as tourism websites and apps. This is an ambitious yet promising direction for scaling up language understanding to richer domains. There is no denying in saying that machine translation plays a pivotal role in this situation. Therefore, it is high time that we should stress the significance of machine translation.

Named Entity Recognition (NER) is defined as finding names in an open do-main text and classifying them among several predefined categories, such as the person, location, and organization names [3]. It not only is the fundamental task of Natural Language Processing (NLP), but also the basic work on machine translation. In addition, it is a very important step in developing other downstream NLP applications [3]. More importantly, it also plays an indispensable role in other natural language processing tasks, such as information extraction, information retrieval, knowledge mapping, knowledge map, question answering system and so on. Therefore, this is a very challenging problem in the field of natural language processing (NLP).

In recent years, Bidirectional Encoder Representation from Transformers (BERT) has performed extremely well in multiple tasks in the field of natural language

© Springer Nature Singapore Pte Ltd. 2020
J. Li and A. Way (Eds.): CCMT 2020, CCIS 1328, pp. 11–23, 2020.
https://doi.org/10.1007/978-981-33-6162-1_2

processing. Most open source monolingual BERT models support English or Chinese, but none support Mongolian. For this purpose, we proposed a BERT pre-training language model suitable for Mongolian researchers, and trained a NER model based on the BERT tagging model by using our Mongolian tourism labeling corpus (Fig. 1).

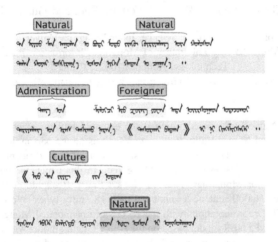

Fig. 1. Examples of Mongolian travel-related named entities in a Mongolian post.

In this paper, we present a comprehensive study to explore the unique challenges of named entity recognition in the tourism field. These named entities are often ambiguous, for example, there is one person name and one location name in the given sentence, the word "ᠣᠣᠣ/" commonly refers to a location name, but also be used as a personal name.

To identify these entities, we propose a NER model based on the Bert-Base tag model, which is a very powerful baseline model that identifies 18 types of travel-related named entities. This model combines local sentence-level context information with remote monitoring information. The NER model is strictly tested by using 16000 newly annotated Mongolian tourism corpus, and its performance is better than the traditional CRF model and BiLSTM-CRF model. Our major contributions are as follows:

A NER Corpus for Mongolian Tourism. 18 types of named entities were manually annotated, including most Mongolian tourism information.

We demonstrate that Named Entity Recognition in the tourism domain is an ideal benchmark task for testing the effectiveness of contextual word representations, such as ELMo and BERT, due to its inherent polysemy and salient reliance on context [1].

A Mongolian NER Model Based on BERT-Base Tagging Model. Eighteen types of fine-grained named entities related to tourism can be identified in the Mongolian Tourism Named Entity Recognition Corpus (MTNER).

Overall, our NER model extracted 18 travel-related named entity types, which scored 82.09% F1 in the Mongolian Tourism Named Entity Recognition Corpus (MTNER). This performance, we believe, is sufficiently strong to be practically useful. And we will release our data and code, including our annotated corpus, annotation guideline, a specially designed tokenizer, and a pre-trained Mongolian BERT and a trained NER model.

2 Related Work

Machine translation has been becoming more and more popular, especially in many special fields, including tourism. At the same time in the field of artificial intelligence knowledge also has great application prospects [7].

The CoNLL 2003 dataset is a widely used bench-mark for the NER task. State-of-the-art approaches on this dataset use a bidirectional LSTM with the conditional random field and contextualized word representations.

Among all the methods, relay on the features to classify the input word, and do not count on linguists to make rules, supervised learning approaches have been the predominant in this filed. In the learning machine, each input will output a label with the learned algorithm, such as Hidden Markov Model [9], Support Vector Machine [10], Conditional Random Fields [11], and so on. Transfer-learning is also used for the NER task [2, 3, 26]. Various methods have been investigated for handling rare entities, for example incorporating external context or approaches that make use of distant supervision.

Named Entity Recognition has been explored for new domains and languages, such as social media, biomedical texts, multilingual texts, and the tourism domain. As to the techniques applied in NER, there are mainly the following streams. At first, rule-based approaches was the mainstream, which do not need annotated data as they rely on hand-crafted rules. Later, unsupervised learning approaches prevailed, which rely on un-supervised algorithms without hand-labeled training examples. Because of feature plays an vital role in the named entity recognition task, and feature-based methods become an inevitable trend, which rely on supervised learning algorithms with careful feature engineering. In recent years, with the development of deep learning, the method based on deep learning has become the mainstream, which automatically discover representations needed for the classification and detection from raw input in an end-to-end manner [25].

In-domain research, the formerly Named Entity Recognition relay on feature engineerings, such as CRF and CRF-CRF [7, 16] be used in the tourism domain, and much Statistical learning also uses into the tourism Named Entity Recognition, including HMM [17]. He latest research, the BERT-BiLSTM-CRF [7] be used in the Chinese military domain and got an excellent result. o we trained a NER model based on BERT-base tagging model for Mongolian in the tourism domain. n linguistics, a corpus or text corpus is a large and structured set of texts, and nowadays usually electronically stored and processed. What's more, in corpus linguistics, corpus is used to do statistical analysis and hypothesis testing, checking occurrences, or validating linguistic rules within a specific language territory [20, 24]. Consequently, it is crucial to build a quantity and quality corpus.

There has been relatively little prior work on named entity recognition in the tourism domain, use BERT-BiLSTM-CRF in Chinese tourism named entity recognition [19]. In this paper, we collected vast Mongolian data to pre-train a pre-training language model for Mongolian, built a corpus for Mongolian Tourism corpus, and annotated 18 types of travel-related entities to train a NER model base on the Mongolian Tourism corpus.

3   Challenge for Mongolian Tourism NER

In this section, we discuss the challenge for Mongolian language understanding and named entity recognition in tourism domain.

The named entities in the general domain, mainly including the names of the person, location, and organization name, have the characteristics of relatively stable type, standardized structure, and unified naming rules. While, in the tourism domain, named entities not only have the general domain challenges, including Large scale vocabulary, lack of abundant corpus, absence of capital letters in the orthography, multi-category word, subject-object-verb word order but also face other in-domain challenges, such as the entity boundary is not clear, the simplification expression, the rich entity types, the large quantity and so on.

The Simplification Expression. For example, "ᠮᠣ ᠠᠠᠮ ᠠᠠᠮᠡ" (Inner Mongolia University) also is said to "ᠠᠠᠠᠠ ᠮᠣ". Those phenomena increase the difficulty of identification.

The Rich Entity Types. In the tourism domain, have many category entities, such as the display name in the scenic spot, is also the named entity should be annotated. It leads to many travel-related named entity need to recognize.

The Large Quantity. Various types named entity in the tourism domain, make the data quantity is large.

The research of NER in Mongolian started late and there are few related works, which largely restricted the development of informatization and intelligentization of Mongolian. In these years, NER has been emerging in the research on Mongolian language information processing. Significant achievements have been made in the identification of three categories of entities, person, location, and organization name. However, few actual achievements could have been made in the research on NER in other specific fields, including the tourism field.

In this paper, we collected a lot of tourism Mongolian data, and build an in-domain Mongolian tourism corpus for named entity recognition, meanwhile, we trained a Mongolian NER model based on BERT-base tagging model.

4 Annotated Mongolian Tourism Corpus

In this section, we describe the construction of our annotated Mongolian tourism NER corpus (MTNER). We collected Mongolian text data, and manually annotated them with 18 types of travel-related entities.

4.1 Data Collection

We collected Mongolian datasets, large and various, about 10 GB, such as Mongolian news and Mongolian tourism, from many websites, including the Mongolian News website of China (http://www.nmgnews.com.cn/), Holovv (https://holoov.com), Ctrip (https://vacations.ctrip.com) and so on. Original datasets genres and sentences number in Table 1.

Table 1. Original datasets genres and sentences number.

Data genres	Sentence number
News	24,593
Essays	256,959
Scenic Spot Intros	2,887
Travel notes	2,657
Others	1,744,681
Total	2,031,777

4.2 Annotation Schema

Based on the investigation and analysis, combined with the characteristic of the tourism domain, we find the traditional three categories, person name, location name, and organization name, is not enough, such as the location of tourism including the general location and scenic spot [7, 16, 17, 21].

Above all, we defined and annotated 18 types of fine-grained entities, including 2 types of Person entities, 6 types of Scenic entities, 4 tapes of Cultural entities, 4 types of Organization entities, and 2 types of Specific Field entities. The Person entities include mentions of Mongolian and Foreigner. The Scenic entities include mentions of Administration place, Natural sight, Public building, Marker building, Business, and Religion place. The Cultural entities include mentions of Culture, Education, Sports, and Musical production. The Organization entities include mentions of Company, Politics, and Charity. What's more, the Specific Field entities include mentions of Military, and Car, in Table 2.

Table 2. Annotated entity classes and examples.

Coarse-grained	Fine-grained	Example	Means
Person	Mongolian	ᠲᠣᠷᠤᠢ	Gentle
	Foreigner	ᠵᠠᠺ ᠵᠠᠺ	Jack
Scenic	Administration place	ᠬᠥᠬᠡᠬᠣᠲᠠ	Hohhot
	Natural sight	ᠶᠡᠬᠡ ᠴᠢᠩ ᠠᠭᠤᠯᠠ	the Big Qing Mountain
	Public place	ᠴᠠᠭᠠᠨ ᠲᠣᠪᠳᠣ᠋ ᠬᠣᠯᠣ ᠨᠢᠰᠬᠡᠯ	Baita International Airport
	Marker building	ᠬᠥᠩᠬᠡ ᠬᠡᠷᠡᠮ	the Drum Tower
	Business	ᠸᠠᠨ ᠳᠠ ᠲᠠᠯᠠᠪᠠᠢ ᠲᠠᠯᠠᠪᠠᠢ	the Wanda Squre
Cultural	Religion	ᠵᠣᠬᠠᠩ ᠰᠦᠮᠡ ᠰᠦᠮᠡ	Jokhang Temple
	Culture	ᠲᠢᠶᠠᠨ ᠪᠢᠶᠠᠨ	Tianbian
	Education	ᠮᠣᠩᠭᠣᠯ ᠤᠨ ᠤᠯᠠᠮᠵᠢᠯᠠᠯᠲᠤ ᠰᠣᠶᠣᠯ	the Mongolian Traditional Culture
	Sports	ᠪᠥᠬᠡ	Wrestling
	Music	ᠯᠢᠮᠪᠠ	Flute
Organizations	Department	ᠥᠪᠥᠷ ᠮᠣᠩᠭᠣᠯ ᠤᠨ ᠶᠡᠬᠡ ᠰᠤᠷᠭᠠᠭᠤᠯᠢ	Inner Mongolia University
	Company	ᠬᠤᠸᠠ ᠸᠧᠢ ᠺᠣᠮᠫᠠᠨᠢ	HUAWEI
	Politics	ᠺᠣᠮᠫᠢᠦᠲᠧᠷ ᠤᠨ ᠨᠡᠶᠢᠭᠡᠮ	ACM
	Charity	ᠤᠯᠠᠭᠠᠨ ᠵᠠᠭᠠᠯᠮᠠᠢ ᠶᠢᠨ ᠨᠡᠶᠢᠭᠡᠮ	the Red Cross Society
Other Fields	Military	ᠲᠠᠩᠺ	Tank
	Car	ᠪᠧᠨᠼ	Mercedes-Benz

We adopt BIOES Label schema, "B" represents the starting position named entity, "I" is inner a named entity, "E" means the ending position named entity, the single entity will be labeled "S". While, others will be labeled "O". That is all, we annotated 73 types of labels.

4.3 Annotation Agreement

Our corpus was annotated by eight annotators, who are college students, majored in computer science, are Mongols.

We used a web-based annotation tool, BRAT, and provided annotators with links to the original travel-related datasets of our collections. We adopted the cross-annotating strategy, four steps following:

1. We divide the data into eight parts and divided annotators into four groups.
2. Everyone annotated one part data.

3. Members of the same group should exchange data to cross-annotate. The inter-annotator agreement is 0.85, measured by span-level Cohen's Kappa (Cohen, 1960).
4. Manually check the marked results.

After those annotated operates, we got the annotated results, those can be saved to the.ann files, including four columns, ID, entity type, start position and end position, entity, in Fig. 2.

T1	Education	3491	3509	《 ᠤᠰᠤᠨ ᠪᠠᠩᠬᠢ 》 ᠨ	(《 Water bank》)
T2	Administration	2159	2176	ᠢᠲ᠋ᠠᠯᠢ ᠶᠢᠨ ᠤᠯᠤᠰ	(Italy)
T3	Foreigner	2388	2396	ᠰᠢᠯᠥᠬᠣ	(Mr. Shiloh)
T4	Foreigner	2572	2580	ᠰᠢᠯᠥᠬᠣ	(Mr. Shiloh)
T5	Foreigner	2696	2704	ᠰᠢᠯᠥᠬᠣ	(Mr. Shiloh)
T6	Mongolian	2762	2772	ᠵᠢᠯ · ᠸᠠᠲ᠋ᠰ	(Watts, Jill)
T7	Mongolian	3472	3490	ᠨᠠ · ᠰᠣᠨᠲ᠋ᠢᠶᠠᠨ ᠶᠢᠨ	(Nathan Sontiian)
T8	Mongolian	3716	3734	ᠨᠠ · ᠰᠣᠨᠲ᠋ᠢᠶᠠᠨ ᠶᠢᠨ	(Nathan Sontiian)
T9	Education	3735	3748	《 ᠤᠰᠤᠨ ᠪᠠᠩᠬᠢ 》	(《 Water bank》)
T10	Mongolian	4624	4640	ᠳᠠᠭᠠᠮᠠᠯᠯᠠᠭᠰᠠᠨ ᠶᠢᠨ	(Dag admiral)

Fig. 2. Annotated results, including ID, entity type, start position, end position, and entity. And the right-most column which we give the English translation of Mongolian.

5 Mongolian Tourism NER Model

In this section, we introduce our Mongolian Tourism NER model with pre-training and fine-tuning strategy. We pre-trained BERT model for Mongolian, and fine-tuned a NER model based on BERT-base tagging model.

BERT, Bidirectional Encoder Representations from Transformers, is a new method of pre-training language representations which obtains state-of-the-art results on a wide array of Natural Language Processing (NLP) tasks. One important aspect of BERT is that it can be adapted to many types of NLP tasks very easily [5].

Pre-trained Language Model. We use collected Mongolian corpus, unlabeled, about 10 GB, releasing the BERT-base, pre-trained a BERT model for Mongolian (Mongolian_base (12-layer, 768-hidden, 12-heads)) [12]. Training parameters in Table 4.

Mongolian Tourism NER Model. The corpus was converted into BIOES label schema for fine-tuning the BERT-base model [12]. We trained our classifier task for NER base on the Mongolian Tourism corpus [13]. Training parameters in Table 3. Model structure in Fig. 3.

Table 3. Pre-training, Fine-tuning parameters, and values.

	Parameter	Value
Pre-training	max_sequence_length	128
	max_predictions_per_seq	20
	masked_lm_prob	0.15
	train_batch_size	8
	num_train_steps	200,000
	learning_rate	2e–5
Fine-tunning	train_batch_size	8
	eval_batch_size	8
	predict_batch_size	8
	learning_rate	5e–5
	num_train_epochs	3
	max_seq_length	128

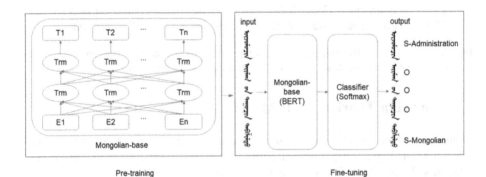

Fig. 3. NER model fine-tuned the BERT-base tagging model. Trained our NER model base on our Mongolian Tourism Named Entity Recognition corpus. Input a sentence, and output including each word label. Such as, in the sentence "ᠤᠤᠵᠤᠪᠠ ᠮᠢᠨᠢ ᠤ᠋ ᠡᠮᠤᠨ᠎ᠡ ᠳᠠᠪᠤᠰᠢᠯᠠᠲᠤ", the word "ᠤᠤᠵᠤᠪᠠ " means "WuZhu MuQin", is a Location name and it only has one word which we annotated "S-Administration" tag, and the word "ᠳᠠᠪᠤᠰᠢᠯᠠᠲᠤ" means "Dabuxilatu", is a Mongolian name and it only has one word which we annotated the "S-Mongolian" tag.

6   Experiment

In this section, we show that our Mongolian NER model outperformance. To evaluate the Mongolian tourism domain named entity recognition model proposed in this paper, fine-tuned the BERT-base tagging model base on Mongolian Tourism Corpus named entity recognition (MTNER), and compared with those mainstream models for named entity recognition work with our corpus, including CRF and BiLSTM-CRF.

6.1   Data

The original Mongolian text data which we collected, have various problems, such as misspelling problem, we use the spelling correction to solve those errors [15], got the unannotated Mongolian corpus and the annotated Mongolian Tourism corpus for NER.

Mongolian Corpus. We pre-trained a BERT model for Mongolian base on our Mongolian corpus, unannotated, about 10 GB. We divided the data into three parts, train, development, and test set.

Mongolian Tourism Corpus for NER. We train and evaluate our NER model on the Mongolian Tourism corpus of 12,800 train, 1,600 development, and 1,600 test sentences. We used the manually annotated corpus in (§4), it is a manually annotated Mongolian Tourism corpus, it contains 16,000 sentences and 15,320 named entities. The person, scenic, cultural, organization, and other fields named entities account for 22.56%, 32.53%, 15.25%, 20.36%, and 9.30%. The account of the fine-grained class in Table 4.

Table 4. The account of fine-grained class.

Entity type	Proportion (%)	Entity type	Proportion (%)
Mongolian	14.30	Education	4.62
Foreigner	8.26	Sports	2.37
Administration place	4.78	Music	2.63
Natural sight	5.74	Department	8.40
Public place	4.17	Company	3.26
Marker building	5.28	Politics	4.23
Business	4.25	Charity	4.47
Religion	8.30	Military	3.60
Culture	5.63	Car	5.70

6.2   Baselines

We compared our NER model with two mainstream named entity recognition models, our NER model outperformed than them.

A Feature-Based Linear CRF Model. This model uses standard orthographic, contextual, and gazetteer features. We implemented a CRF baseline model to extract the travel-related entities with the word-level input embedding. The regular expressions are developed to recognize specific categories of travel-related entities [11]. We use the "LBFGS" algorithm, and the cost parameters is 0.1.

A BiLSTM-CRF Model. This model uses a BiLSTM-CRF network to predict the entity type of each word from its weighted representations. Using the contextual word embedding (ELMo) embeddings, and is used as the state-of-the-art baseline named entity recognition models in various special domains [12]. We set the word embedding size is 128, and the dimensionality of LSTM hidden states is 128. And set the same initial learning rate, batch size and epochs.

6.3   Results

We evaluated the results by the CoNLL metrics of precision, recall, and F_1 [6]. Precision is the percentage of corrected named entities, recall is the percentage of named entities existing in the corpus and F_1 is the harmonic mean of precision and recall, these can be expressed as:

$$precision = \frac{Num(\text{correct NEs predicted})}{Num(NEspredicted)}$$

$$recall = \frac{Num(\text{correct NEs predicted})}{(\text{Num all NEs})}$$

$$F1 = \frac{2 * precision*recall}{precision + recall}$$

On the same training set and test set, we compared the above two models. Table 5 shows the precision (**P**), recall (**R**), and F_1 score comparison of the different models on our Mongolian Tourism corpus named entity recognition.

Table 5. Evaluation of the test sets of the Mongolian tourism NER corpus.

	P (%)	R (%)	F_1 (%)
CRF	75.10	82.33	78.55
BiLSTM-CRF	76.32	84.25	80.08
Mongolian Tourism NER model	**78.59**	**85.94**	**82.09**

The results indicate that our Mongolian Tourism NER model is better than the other two named entity recognition models. Compared with the CRF named entity recognition model, BiLSTM can learn more contextual features. The model proposed in this paper improves the F_1 by 3.54% and the recall by 3.61%. Compared with BiLSTM-CRF named entity recognition model, our model improves the F_1 by 2.01% and the recall by 1.69%.

The features of word-level ignored the feature with the contextual, this model is a combination of words, sentences, and location features generated word representation, and using the Transformer to train the model, fully considering the influence of the contextual information of the entity, and got a better result.

6.4   Analysis

Pre-trained Language Model. Pre-training on large text corpora can learn common language representations and help complete subsequent tasks, s pre-training is an essential task in NLP. Pre-trained representations can also either be context-free or contextual, and contextual representations can further be unidirectional or bidirectional. Context-free models such as Word2vec, generate a single "word embedding" representation for each word in the vocabulary. Contextual models instead generate a representation of each word that is based on the other words in the sentence, such as ELMO, but crucially these models are all unidirectional or shallowly bidirectional. This means that each word is only contextualized using the words to its left (or right). Some previous work does combine the representations from separate left-context and right-context models, but only in a "shallow" manner. BERT represents one word using both its left and the right context, starting from the very bottom of a deep neural network, so it is deeply bidirectional. BERT outperforms previous methods because it is the first unsupervised, deeply bidirectional system for pre-training NLP [5].

Training Data Scale. Usually, trains a pre-trained language model needs a large corpus. The corpus size of model training directly affects the performance of the model. The large scale of training data enables the model to fully learn the characteristics of language, to make full use of the corpus information to solve the problem of language understanding [20, 21]. So our data scale is not enough, we need to annotate more tourism and another domain corpus to support the downstream NLP task.

The Proportion of Entity Categories. The text classification task, category distribution balance is very important to the classification model. Unbalanced classification makes it easy for the model to forget the categories that appear less frequently [2, 3, 18, 19, 21]. In our Mongolian Tourism corpus, the proportion of annotated entity types is balanced, it could trained a classifier model to be better.

7 Conclusion

In this work, we investigated the task of named entity recognition in the Mongolian Tourism domain. We collected a vast Mongolian text, developed a Mongolian Tourism Corpus of 16,000 sentences from the Mongolian Tourism domain annotated with 18 fine-grained named entities. This new corpus is an benchmark dataset for contextual word representations. We also pre-trained a BERT model for Mongolian and fine-tuned a NER model based on BERT-base tagging model for Mongolian Tourism named entity recognition. This NER model outperforms other mainstream NER models on this dataset. Our pre-trained Mongolian-base consistently helps to improve the Mongolian NER performance. We believe our corpus, BERT embedding for Mongolian, fine-tuned BERT-base tagging model for Mongolian Tourism NER model will be useful for various Tourism tasks and other Mongolian NLP tasks, such as Tourism Knowledge Graph, Mongolian Machine Translation, Mongolian question-answering, and so on.

Acknowledgement. The project are supported by National Natural Science Foundation of China (No. 61773224); Inner Mongolia Science and Technology Plan (Nos. 2018YFE0122900, CGZH2018125, 2019GG372, 2020GG0046); Natural Science Foundation of Inner Mongolia (Nos. 2018MS06006, 2020BS06001); Research Fund for Inner Mongolian Colleges (No. NJZY20008); Research Fund for Inner Mongolian Returned Oversea Students and Inner Mongolia University Outstanding Young Talents Fund.

References

1. Tabassum, J., Maddela, M., Xu, W., et al.: Code and Named Entity Recognition in StackOverflow. arXiv (2020)
2. Wang, W, Bao, F., Gao, G.: Learning morpheme representation for mongolian named entity recognition. Neural Process. Lett **50**, 2647–2664 (2019)
3. Wang, W, Bao, F., Gao, G.: Mongolian named entity recognition with bidirectional recurrent neural networks. In: The 28th IEEE International Conference on Tools with Artificial Intelligence (ICTAI 2016), pp. 495–500 (2016)
4. Marcus, M.P., Marcinkiewicz, M.A., Santorini, B., et al.: Building a large annotated corpus of English: the penn treebank. Comput. Linguist. **19**(2), 313–330 (1993)
5. Devlin, J., Chang, M., Lee, K., et al.: BERT: pre-training of deep bidirectional transformers for language understanding. In: North American chapter of the Association for Computational Linguistics, pp. 4171–4186 (2019)
6. Nadeau, D., Sekine, S. A survey of named entity recognition and classification. Lingvae Investigationes. **30**(1), 3–26 (2007)
7. Geng, X.: Research and Construction of the Map of Mongolian and Chinese Bilingual Knowledge for Tourism (2019)
8. Cao, Y., Hu, Z., Chua, T., et al.: Low-resource name tagging learned with weakly labeled data. In: International Joint Conference on Natural Language Processing, pp. 261–270 (2019)
9. Zhou, G., Named entity recognition using an HMM-based chunk tagger. In: Proceedings of North American chapter of the Association for Computational Linguistics 2002, pp. 473–480 (2002)
10. Kudo, T., Matsumoto, Y.: Chunking with support vector machines. North American chapter of the Association for Computational Linguistics, 1508.01991 (2001)
11. Lafferty, J., Mccallum, A., Pereira, F.: Conditional random fields: probabilistic models for segmenting and labeling sequence data. In: Proceedings of 18th International Conference on Machine Learning (ICML), pp. 282–289 (2002)
12. Huang, Z., Xu, W., Yu, K.: Bidirectional LSTM-CRF models for sequence tagging. Comput. Sci. (2015)
13. Sun, C., Qiu, X., Xu, Y., Huang, X.: How to fine-tune BERT for text classification? In: Sun, M., Huang, X., Ji, H., Liu, Z., Liu, Y. (eds.) CCL 2019. LNCS (LNAI), vol. 11856, pp. 194–206. Springer, Cham (2019). https://doi.org/10.1007/978-3-030-32381-3_16
14. Yin, X., Zhao, H., Zhao, J., Yao, W., Huang, Z.: Named entity recognition in military field by multi-neural network collaboration. J. Tsinghua Univ. **60**(08), 648–655 (2020)
15. Lu, M., Bao, F., Gao, G., Wang, W., Zhang, H.: An automatic spelling correction method for classical mongolian. In: Douligeris, C., Karagiannis, D., Apostolou, D. (eds.) KSEM 2019. LNCS (LNAI), vol. 11776, pp. 201–214. Springer, Cham (2019). https://doi.org/10.1007/978-3-030-29563-9_19

16. Guo, J., Xue, Z., Yu, Z., et al.: Named entity identification in tourism based on cascading conditions. Chinese J. Inf. Technol. **023**(005), 47–52 (2009)
17. Xue, Z., Guo, J., Yu, Z., et al.: Identification of Chinese tourist attractions based on HMM. J. Kunming Univ. Sci. Technol. **34**(006), 44–48 (2009)
18. Dongdong, L.: Named entity recognition for medical field (2018)
19. Zhao, P., Sun, L., Wan, Y., Ge, N.: BERT + BiLSTM + CRF based named entity recognition of scenic spots in Chinese. Comput. Syst. Appl. **29**(06), 169–174 (2020)
20. Wang, C.: The Research and construction of Yi corpus for information processing. Int. J. New Dev. Eng. Soc. **3**(4), 57–63 (2019)
21. Lin, B., Yip, P.C.: On the construction and application of a platform-based corpus in tourism translation teaching. Int. J. Translation Interpretation Appl. Linguist. **2**(2), 30–41 (2020)
22. Ren, Z., Hou, H., Jia, T., Wu, Z., Bai, T., Lei, Y.: Application of particle size segmentation in the translation of mongolian and Chinese neural machines. Chinese J. Inf. Technol. **33** (01), 85–92 (2019)
23. Cui, J., Zheng, D., Wang, D., Li, T.: Entity recognition for chrysanthemum named poems based on deep learning model. Information Theory and Practice pp. 1–11 (2020)
24. Liu, G.: Construction of parallel corpus for legal translation. Overseas English. (10) 32–33 (2020)
25. Li, J., Sun, A., Han, J., et al.: A survey on deep learning for named entity recognition. IEEE Trans. Knowl. Data Eng. 1. (2020)
26. Wang, W., Bao, F., Gao, G.: Mongolian named entity recognition system with rich features. In: Proceedings of COLING 2016, the 26th International Conference on Computational Linguistics: Technical Papers, pp. 505–512 (2016)

Unsupervised Machine Translation Quality Estimation in Black-Box Setting

Hui Huang[1], Hui Di[2], Jin'an Xu[1(✉)], Kazushige Ouchi[2],
and Yufeng Chen[1]

[1] Beijing Jiaotong University, Beijing, China
{18112023, jaxu, chenyf}@bjtu.edu.cn
[2] Toshiba (China) Co., Ltd, Beijing, China
{dihui, kazushige.ouchi}@toshiba.com.cn

Abstract. Machine translation quality estimation (Quality Estimation, QE) aims to evaluate the quality of machine translation automatically without golden reference. QE is an important component in making machine translation useful in real-world applications. Existing approaches require large amounts of expert annotated data. Recently, there are some trials to perform QE in an unsupervised manner, but these methods are based on glass-box features which demands probation inside the machine translation system. In this paper, we propose a new paradigm to perform unsupervised QE in black-box setting, without relying on human-annotated data or model-related features. We create pseudo-data based on Machine Translation Evaluation (MTE) metrics from existing machine translation parallel dataset, and the data are used to fine-tune multilingual pre-trained language models to fit human evaluation. Experiment results show that our model surpasses the previous unsupervised methods by a large margin, and achieve state-of-the-art results on MLQE Dataset.

Keywords: Machine translation · Unsupervised quality estimation · Pre-trained language model

1 Introduction

In recent years, with the development of deep learning, Machine Translation (MT) systems made a few major breakthroughs and were wildly applied. Machine translation quality estimation (Quality Estimation, QE) aims to evaluate the quality of machine translation automatically without golden reference [1]. The quality can be measured with different metrics, such as HTER (Human-targeted Edit Error) [2] or DA (Direct Assessment) Score [3].

Previous methods treat QE as a supervised problem, and they require large amounts of in-domain translations annotated with quality labels for training [4, 5]. However, such large collections of data are only available for a small set of languages in limited domains.

H. Huang—Work was done when Hui Huang was an intern at Research and Develop Center, Toshiba (China) Co., Ltd., China.

Recently, Fomicheva [6] firstly performs QE in an unsupervised manner. They explore different information that can be extracted from the MT system as a by-product of translation, and use them to fit quality estimation output. Since their methods are based on glass-box features, they can only be implemented in limited situations and demands probation inside the machine translation system.

In this work, we firstly propose to perform unsupervised QE in a black-box setting, without relying on human-annotated data or model-related features. We create pseudo-data based on Machine Translation Evaluation (MTE) metrics, such as BLEU, HTER and BERTscore, from publicly-accessible translation parallel dataset. The MTE-metrics based data are then used to fine-tune several multilingual pre-trained language models, to evaluate translation output.

To the best of our knowledge, this is the first work to utilize MTE methods to deal with QE. Our method does not involve complex architecture engineering and easy to implement. We performed experiment on two language-pairs on MLQE[1] Dataset, outperforming Fomicheva by a large margin. We even outperformed two supervised models of Fomicheva, revealing the potential of MTE-based methods for QE.

2 Background

2.1 Machine Translation Evaluation

Similar to QE, Machine Translation Evaluation (MTE) also aims to evaluate the machine translation output. The difference between MTE and QE is that MTE normally requires annotated references, while QE is performed without reference and highly relies on source sentences.

Human evaluation is often the best indicator of the quality of a system. However, designing crowd sourcing experiments such as Direct Assessment (DA) [3] is an expensive and high-latency process, which does not easily fit in a daily model development pipeline.

Meanwhile automatic metrics, for example BLEU [7] or TER [2], can automatically provide an acceptable proxy for quality based on string matching or hand-crafted rules, and have been used in various scenarios and led the development of machine translation. But these metrics cannot appropriately reward semantic or syntactic variations of a given reference [8].

Recently, after the emergence of pre-trained language models, a few contextual embedding based metrics have been proposed, such as BERTscore [8] and BLUERT [9]. These metrics compute a similarity score for the candidate sentence with the reference based on token embeddings provided by pre-trained models. Refraining from relying on shallow string matching and incorporate lexical synonymy, BERTscore can achiev higher relevance with human evaluation.

Given the intrinsic correlation nature of MTE and QE, few works have been done to leverage MTE methods to deal with the task of QE.

---

[1] https://github.com/facebookresearch/mlqe.

2.2 Machine Translation Quality Estimation

Despite the performance of machine translation systems is usually evaluated by automatic metrics based on references, there are many scenarios where golden reference is unavailable or hard to get. Besides, reference-based metrics also completely ignore the source segment [10]. This leads to pervasive interest on the research of QE.

Early methods referred to QE as a machine learning problem [11]. Their model could be divided into the feature extraction module and the classification module. Highly relied on heuristic artificial feature designing, these methods did not manage to provide reliable estimation results.

During the trending of deep learning in the field of natural language processing, there were also a few works aiming to integrate deep neural network into QE systems. Kim [12] proposed for the first time to leverage massive parallel machine translation data to improve QE results. They applied RNN-based machine translation model to extract high-quality feature. Fan [13] replaced the RNN-based MT model with Transformer and achieved strong performance.

After the emergence of BERT, there were a few works to leverage pretrained models on the task of QE [14, 15]. Language models pre-trained on large amounts of text documents are suitable for data-scarce QE task by nature, and have led to significant improvements without complex architecture engineering.

Despite most models relied on artificial annotated data, there were also a few trials aiming to apply QE in an unsupervised manner. The most important work is Fomicheva [6], which proposed to fit human DA scores with three categories of model-related features: A set of unsupervised quality indicators that can be produced as a by-product of MT decoding; the attention distribution inside the Transformer architecture; model uncertainty quantification captured by Monte Carlo dropout. Since these methods are all based on glass-box features, they can only be applied in limited scenarios where inner exploration into the MT model is possible.

3 Model Description

3.1 Pretrained Models for Quality Estimation

Our QE predictor is based on three different pre-trained models, namely BERT [16], XLM [17], and XLM-R [18], as shown in Fig. 1.

Given one source sentence and its translated result, our model concatenates them and feeds them into the pre-trained encoder. To leverage the global contextual information when doing sentence-level prediction, an extra layer of bidirectional recurrent neural network is applied on the top of the pre-trained model.

Despite the shared multilingual vocabulary, BERT is originally a monolingual model [19], pretrained with sentence-pairs from one language or another. To help BERT adapts to our bilingual scenario, where the inputs are two sentences from different languages, we implement a further pre-training step.

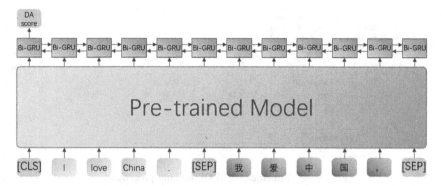

Fig. 1. Pre-trained model for quality estimation.

During the further pre-training step, we combine bilingual sentence pairs from large-scale parallel dataset, and randomly mask sub-word units with a special token, and then train BERT model to predict masked tokens. Since our input are two parallel sentences, during the predicting of masked words given its context and translation reference, BERT can capture the lexical alignment and semantic relevance between two languages.

In contrast, XLM and XLM-R are multilingual models by nature, which receive two sentences from different languages as input during training, that means a further pre-training step is redundant. The training strategies and data of XLM and XLM-R are designed distinctly, which are explained in detail in their papers.

3.2 MTE-Based QE Data

Despite sentence-pairs with source and machine-translated text readily accessible (for which we only need to translate source text into target language using a MT system), the absence of DA scores becomes our biggest challenge. Even in supervised scenario, human-annotated DA scores are still scarce and limited [5]. Therefore, we propose to use MTE metrics to fit human assessment, thus creating massive pseudo data for the training of the QE system. Our approach can be described as follows:

Firstly, we decode source sentences in parallel corpus into target language. Secondly, we use automatic MTE-metrics to evaluate the quality of output sentences based on references. In this step we do not need any human annotation or time-consuming training. The MTE based evaluation can give a roughly accurate quality assessment, and can be used as substitution to human-annotated DA scores. Thirdly, the pseudo DA scores, combined with source and translated sentence pairs, are used to train our QE system (Fig. 2).

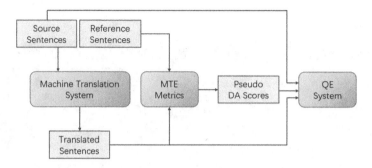

Fig. 2. MTE-metrics based QE training procedure.

We tried three different MTE metrics to fit DA evaluation, namely TER [2], BLEU [7], and BERTscore [8].

TER uses word edit distance to quantify similarity, based on the number of edit operations required to get from the candidate to the reference, and then normalizes edit distance by the number of reference words, as shown in Eq. 1.

$$TER = \frac{\# \; of \; edits}{average \; \# \; of \; reference \; words} \tag{1}$$

BLEU is the most widely used metric in machine translation. It counts the number of n-grams that occur in the reference sentence and candidate sentence. Each n-gram in the reference can be matched at most once, and very short candidates are discouraged using a brevity penalty, as shown in Eq. 2.

$$BLEU = BP \cdot exp\left(\sum_{n=1}^{N} w_n \log p_n\right), BP = \begin{cases} 1 & if \; c > r \\ e^{(1-r/c)} & if \; c \le r \end{cases} \tag{2}$$

where p_n denotes the geometric average of the modified n-gram precisions, w_n denotes positive weight for each token, c denotes the length of the candidate translation and r denotes the effective reference corpus length.

BERTscore calculates the cosine similarity of a reference token and a candidate token based on their contextual embedding provided by the pre-trained model. The complete score matches each token in reference to a token in candidate to compute recall, and each token in candidate to a token in reference to compute precision, and then combine precision and recall to compute an F1 measure, as displayed in the following equations.

$$P_{BERT} = \frac{1}{|\hat{x}|} \sum_{\hat{x}_j \in \hat{x}} \max_{x_i \in x} x_i^T \hat{x}_j \tag{3}$$

$$R_{BERT} = \frac{1}{|x|} \sum_{x_i \in x} \max_{\hat{x}_j \in \hat{x}} x_i^T \hat{x}_j \tag{4}$$

$$F_{BERT} = 2 \frac{P_{BERT} \cdot R_{BERT}}{P_{BERT} + R_{BERT}} \tag{5}$$

where x and $\hat{x}$ denote the contextual embedding for each token in reference and candidate sentences, respectively.

With the ability of matching paraphrases and capturing distant dependencies and ordering, BERTscore is proved to be highly correlated with human assessment [8].

4   Experiment

4.1   Setup

Dataset. The dataset we use is MLQE Dataset [6], which contains training and development data for six different language-pairs. We performed our experiments mainly on two high-resource languages (English–Chinese and English–German). Since we want to solve the problem in unsupervised setting, we only used the 1000 sentence-pairs from the development data for each direction respectively.

To train our own MT model, we use the WMT2020 English-Chinese and English-German data[2], which contains roughly 10 million sentence-pairs for each direction after cleaning (a large proportion is reserved to generate QE data).

Fomicheva also provide the MT model which was used to generate their QE sentence pairs, thus we have two different MT models to use. We will explain the influence of different MT models in the next section.

For fine-tuning pre-trained models, we used the reserved data from WMT2020 English-Chinese and English-German translation, and randomly sampled 500 k sentence pairs for each direction to create MTE-based QE data.

Baseline. Sine there are few works done in the area of unsupervised QE, we mainly make comparison with Fomicheva. They proposed 10 methods which can be categorized as three sets, among them we display their top-two results in each direction, namely D-Lex-Sim and D-TP for English-Chinese, and D-TP and Sent-Std for English-German.

We also make comparison with supervised methods, including PredEst models using the same parameters in the default configurations provided by Kepler [14], and the recent SOTA QE system BERT, augmented with two bidirectional RNN [15]. These two models are trained with the provided 7000 training pairs.

4.2   Experiment Results

As shown in Table 1 and Table 2, our approach surpasses Fomicheva with their best-performance methods by a large margin on both directions, verifying the effectiveness

---

[2] http://www.statmt.org/wmt20/translation-task.html.

Table 1. Experiment results on English-Chinese MLQE Dataset.

Language direction	Method	Pearsonr	Spearman
English-Chinese	PredEst	0.190	–
	BERT-BiRNN	0.371	–
	D-Lex-Sim	0.313	–
	D-TP	0.321	–
	TER	0.3919	0.4116
	BLEU	0.3668	0.3941
	BERTscore-precision	0.4254	0.4347
	BERTscore-F1	**0.4288**	0.4373

Table 2. Experiment results on english-German MLQE dataset.

Language direction	Method	Pearsonr	Spearman
English-German	PredEst	0.145	–
	BERT-BiRNN	0.273	–
	D-TP	0.259	–
	Sent-Std	0.264	–
	TER	0.2589	0.2828
	BLEU	0.2637	0.2931
	BERTscore-precision	**0.3124**	0.3327
	BERTscore-F1	0.3089	0.3284

of MTE-based QE data. We even outperform BERT-BiRNN trained in supervised manner on both directions.

Although the supervised training data provided is limited and our best results are achieved by XLM rather than BERT (we will explain this in next section), the result is still very fascinating.

The glass-box features, although thoroughly explored by Fomicheva, seem unhelpful compared with MTE-metrics based methods. These features are no more than statistic cues regulated by the machine translation model. If we rely on the same MT model to evaluate the translation, then we will be constrained by itself and unable to cope with various phenomena.

Moreover, we can also conclude that when fine-tuning pre-trained models for QE task, the quantity of data is more important than the quality of data, as shown in Fig. 3. Although our data is generated purely based on automatic metrics rather human annotators, we can still surpass supervised systems trained only with clean data.

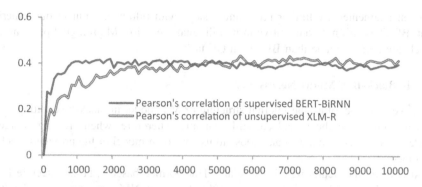

Fig. 3. The variation of Pearson's correlation coefficient with the increase of the training step. Although the supervised model could generate better results in the first few steps, as the unsupervised model receives more data after more steps, it would outperform the supervised model.

Among our three methods, BERTscore-based methods achieve better results than statistical metrics-based methods, which is reasonable since BERTscore is proved to better correlate with human assessment. More accurate MTE metrics could lead to more natural pseudo data, therefore enable the QE model to perform better.

5 Analysis

5.1 Is BERT Always the Best?

Despite the overwhelming results BERT has accomplished on multiple datasets, our scenario demands the ability to process bilingual input, while BERT is originally a monolingual model, treating the input as either being from one language or another.

In contrast, XLM and XLM-R are multilingual models by nature, pre-trained with bilingual inputs from different languages. Since QE task aims to evaluate the translation based on the source sentence from another language, XLM and XLM-R should be more suitable. Experiment results in Table 3 verify our hypothesis.

Table 3. Experiment results on MLQE direct assessment data.

Language direction	Method	Pretrained model	Pearonr	Spearman
English-Chinese	BERTscore-precision	BERT	0.3255	0.3295
		BERT(further-trained)	0.3827	0.3895
		XLM	**0.4254**	0.4347
		XLM-R	0.4170	0.4227
	BERTscore-F1	BERT	0.3271	0.3329
		BERT(further-trained)	0.3836	0.3889
		XLM	0.4110	0.4221
		XLM-R	**0.4288**	0.4373

Even augmented by further pre-training steps with bilingual input in our experiment, BERT is still not competitive in multilingual scenarios. Multilingual pre-trained models are more suitable than BERT on QE task.

5.2   Is Black-Box Model Necessary?

While we cannot explore the internal structure of MT model in black-box setting, the input and output of the model are still available. Therefore, when creating source-translation sentence pairs, we can choose to use our own model or the provided black-box model.

Nowadays, the neural-based (especially Transformed-based) MT architecture has dominated the machine translation area [20]. Different NMT systems trained with similar data may behave similarly to the same input [21].

Therefore, even with another model trained with slightly different data, the generated translation may still have similar error distribution. Experiment results displayed in Table 4 verify our hypothesis.

Table 4. Results of different data generated by different MT models.

Language direction	Method	MT Model	Pearonr	Spearman
English-Chinese	TER	Ours	0.3671	0.3752
		Provided	0.3919	0.4116
	BLEU	Ours	0.3485	0.3619
		Provided	0.3668	0.3941
	BERTscore-precision	Ours	0.3853	0.3998
		Provided	0.4254	0.4227
	BERTscore-F1	Ours	0.3995	0.4133
		Provided	0.4288	0.4373

While the data generated by the provided model does obtain higher correlation, the result obtained by our own model is yet competitive. When creating MTE-based QE data, the provided model can benefit a lot, but if it is not available, we can simulate its error distribution with similar architecture and similar training data.

5.3   Where Is the Limitation of QE?

In this section, we would like to perform a case-study based on our results on development set. Since the distributions of our system's output and the real-world QE scores differ a lot, as shown in Fig. 1, we mainly compare the ranking for the same sentence in different methods. Namely, we would rank the whole development set according to scores provided by our system and the golden label, and compare the discrepancy of ranking for the same sentence in different systems.

In summary, there are two problems impede the performance of our model.

Firstly, our model relies too much on the syntactic consistency while ignoring semantic understandability to evaluate a translation. Given a translated sentence with syntactically consistent structure, our model would assign a very high score even when the translation is semantically erroneous (Fig. 4).

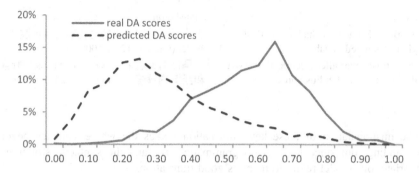

Fig. 4. Distribution of DA scores on development set. Solid line denotes the output of our system, and dashed line denotes the golden labels.

Table 5. Wrong prediction caused by syntactical inconsistency.

Source	Translation 1	Translation 2
A snob, a sneak and a coward, with very few redeeming features.	一个卑鄙的人，一个偷偷摸摸的人，一个懦弱的人，几乎没有什么可取之处。(ranking 993 of 1000)	一个卑鄙，一个偷偷摸摸，一个懦弱，几乎没有什么可取之处。(ranking 837 of 1000)
Others befriended and watched over the peasantry;	另一些人亲密无间地守护着农民 ;(ranking 763 of 1000)	另一些人做朋友并且守护着农民 ;(ranking 631 of 1000)

As shown in Table 5, although Translation 2 is much better than Translation 1, our method would still assign a higher evaluation score for Translation1 since the syntactic structure is more consistent.

This problem originates from pre-trained models themselves, as it is very likely for pre-trained models to rely on spurious statistical cues when doing prediction [22], while not really understand the sentence meaning. Most sentence pairs with a consistent syntactic structure are assigned with a higher score in our training data, which is captured by our model and used as an inappropriate criterion for evaluation.

The second problem is that our system fails to detect erroneously translated words, especially when prior knowledge is in need.

As shown in Table 6, for the first sentence, the provided model mistranslated the word *Judah*, which is a country, as a name. And in the second sentence, the word *consulship*, which refers to a period, is mistranslated as a building. To understand why these words are mistranslated, you may need related history knowledge.

Table 6. Wrong prediction caused by mistranslated words.

Source	Translation
In 586 BCE, King Nebuchadnezzar II of Babylon conquered Judah.	巴比伦国王尼布查德尼扎尔二世征服了犹大. (ranking 12 of 1000)
Roman satirists ever after referred to the year as "the consulship of Julius and Caesar."	罗马讽刺家后来把这一年称为 "朱利叶斯和凯撒的领馆" 。(ranking 225 of 1000)

The mistranslation of these key information makes the whole sentence beyond understanding, but since there is no grammatic error and the syntactic structure is appropriate, our model refers to them as good translations.

For the first problem, we believe it can be alleviated by strategically picked training samples, with more sentence-pairs syntactically inconsistent but semantically correct. We will leave this as our future work.

Since both QE model and MT model are based on deep-learning, QE can barely solve these problems which MT model cannot solve. More training data may help to alleviate this problem, but can hardly solve it, as more training data does not really introduce structured prior knowledge. We believe this is the limitation of QE.

6   Conclusion

Machine translation quality estimation (Quality Estimation, QE) aims to evaluate the quality of machine translation automatically without reference provided. Despite it has attracted a lot of research interest recent years, few works have been done to deal with QE in an unsupervised manner.

In this paper, we have devised an unsupervised approach to QE where we do not rely on any glass-box features. We create massive pseudo data based on automatic machine translation evaluation (MTE) metrics such as BLEU, TER and BERTscore, from publicly accessible machine translation parallel dataset. Then we use the MTE-based QE data to fine-tune multilingual pre-trained models, to predict direct assessment (DA) scores. Our approach surpassed previous unsupervised methods by a large margin, and even surpassed supervised methods, proving the effectiveness of incorporating MTE metrics into QE.

Despite the lack of human-annotated DA scores, the MTE metrics can provide a highly reliable evaluation for machine translated sentences, and enable us to perform QE in an unsupervised way. We will continue to explore the application of MTE in QE models, and try to reach the limitation of deep-learning based QE.

Acknowledgement. This work is supported by the National Natural Science Foundation of China (Contract 61976015, 61976016, 61876198 and 61370130), and the Beijing Municipal Natural Science Foundation (Contract 4172047), and the International Science and Technology Cooperation Program of the Ministry of Science and Technology (K11F100010), and Toshiba (China) Co., Ltd.

References

1. John, B., et al.: Confidence estimation for machine translation. In: Proceedings of the International Conference on Computational Linguistics, p. 315 (2004)
2. Matthew, S., Bonnie, D., Richard, S., Linnea, M., John, M.: A study of translation edit rate with targeted human annotation. In: Proceedings of association for machine translation in the Americas, vol. 200, No. 6 (2006)
3. Yvette, G., Timothy, B., Alistair, M., Justin, Z.: Can machine translation systems be evaluated by the crowd alone. Nat. Lang. Eng. **23**, 1–28 (2015)
4. Hyun, K., Jong-Hyeok, L., Seung-Hoon, N.: Predictor-estimator using multilevel task learning with stack propagation for neural quality estimation. In: Proceedings of the Second Conference on Machine Translation, vol. 2, Shared Tasks Papers, pp. 562–568 (2017)
5. Erick, F., Lisa, Y., André, M., Mark, F., Christian, F.: Findings of the WMT 2019 shared tasks on quality estimation. In: Proceedings of the Fourth Conference on Machine Translation (Shared Task Papers, Day 2), vol. 3, pp. 1–10 (2019)
6. Fomicheva, M., et al.: Unsupervised Quality Estimation for Neural Machine Translation. arXiv preprint arXiv:2005.10608 (2020)
7. Kishore, P., Salim, R., Todd, W., Wei-Jing, Z.: Bleu: a method for automatic evaluation of machine translation. In: Proceedings of the 40th annual meeting of the Association for Computa-tional Linguistics, pp. 311–318 (2002)
8. Zhang, T., Kishore, V., Wu, F., Weinberger, K. Q., Artzi, Y.: Bertscore: evaluating text generation with bert. arXiv preprint arXiv:1904.09675 (2019)
9. Sellam, T., Das, D., Parikh, A.P.: BLEURT: Learning Robust Metrics for Text Generation. arXiv preprint arXiv:2004.04696 (2020)
10. Lucia, S.: Exploiting objective annotations for measuring translation post-editing effort. In: Proceedings of the 15th Conference of the European Association for Machine Translation, pp. 73–80 (2011)
11. Bojar, O., Chatterjee, R., Federmann, C., Graham, Y., Logacheva, V.: Findings of the 2017 conference on machine translation. In: Proceedings of the Second Conference on Machine Translation, pp. 169–214 (2017)
12. Kim, H., Jung, H.-Y., Kwon, H., Lee, J.H., Na, S.-H.: Predictor-estimator: neural quality estimation based on target word prediction for machine translation. ACM Trans. Asian and Low-Resour. Lang. Inf. Proc. (TALLIP) **17**(1), 3 (2017)
13. Kai, F., Bo, L., Fengming, Z., Jiayi W.: "Bilingual Expert" Can Find Translation Errors. arXiv preprint arXiv:1807.09433 (2018)
14. Kepler, F., et al.: Unbabel's Participation in the WMT19 Translation Quality Estimation Shared Task. arXiv preprint arXiv:1907.10352 (2019)
15. Frédéric, B., Nikolaos, A., Lucia, S.: Quality in, quality out: Learning from actual mistakes. In: Proceedings of the 22nd Annual Conference of the European Association for Machine Translation (2020)
16. Devlin, J., Chang, M.W., Lee, K., Toutanova, K.: Bert: pre-training of deep bidirectional transformers for language understanding. arXiv preprint arXiv:1810.04805 (2018)

17. Lample, G., Conneau, A.: Cross-lingual language model pretraining. arXiv preprint arXiv: 1901.07291 (2019)
18. Conneau, A., et al.: Unsupervised cross-lingual representation learning at scale. arXiv preprint arXiv:1911.02116 (2019)
19. Pires, T., Schlinger, E., Garrette, D. How multilingual is Multilingual BERT?. arXiv preprint arXiv:1906.01502 (2019)
20. Barrault, L., et al.: Findings of the 2019 conference on machine translation (wmt19). In: Proceedings of the Fourth Conference on Machine Translation (Shared Task Papers, Day 1), vol. 2, pp. 1–61 (2019)
21. Ma, Q., Wei, J., Bojar, O., Graham, Y.: Results of the WMT19 metrics shared task: Segment-level and strong MT systems pose big challenges. In: Proceedings of the Fourth Conference on Machine Translation (Shared Task Papers, Day 1), vol. 2, pp. 62–90 (2019)
22. Niven, T., Kao, H.Y.: Probing neural network comprehension of natural language arguments. arXiv preprint arXiv:1907.07355 (2019)
23. Tandon, N., Varde, A.S., de Melo, G.: Commonsense knowledge in machine intelligence. ACM SIGMOD Rec. **46**(4), 49–52 (2018)
24. Zhang, J., Liu, Y., Luan, H., Xu, J., Sun, M.: Prior knowledge integration for neural machine translation using posterior regularization. arXiv preprint arXiv:1811.01100 (2018)

YuQ: A Chinese-Uyghur Medical-Domain Neural Machine Translation Dataset Towards Knowledge-Driven

Qing Yu, Zhe Li$^{(\boxtimes)}$, Jiabao Sheng, Jing Sun, and Wushour Slamu

Xinjiang University, Ürümqi, China
yuqing0131@126.com, {lizhe,jiabao}@stu.xju.edu.cn,
{sunjing,wushour}@xju.edu.cn

Abstract. Recent advances of deep learning have been successful in delivering state-of-the-art performance in medical analysis, However, deep neural networks (DNNs) require a large amount of training data with a high-quality annotation which is not available or expensive in the field of the medical domain. The research of medical domain neural machine translation (NMT) is largely limited due to the lack of parallel sentences that consist of medical domain background knowledge annotations. To this end, we propose a Chinese-Uyghur NMT knowledge-driven dataset, **YuQ**, which refers to a ground medical domain knowledge graphs. Our corpus 65K parallel sentences from the medical domain 130K utterances. By introduce medical domain glossary knowledge to the training model, we can win the challenge of low translation accuracy in Chinese-Uyghur machine translation professional terms. We provide several benchmark models. Ablation study results show that the models can be enhanced by introducing domain knowledge.

1 Introduction

Knowledge can improve the translation quality in NMT models where background knowledge plays a vital role in the success of text generation (Li et al. 2016; Shang et al. 2015; Shao et al. 2016). In neural machine translation systems, background knowledge is defined as slot-value pairs, which provide key information for proper noun translation, and has been well defined and thoroughly studied in conversational systems (Wen et al. 2015; Zhou et al. 2016). However, in neural machine translation of terminology, it is important but challenging to leverage background knowledge, which is represented as either knowledge graphs (Zhou et al. 2018a; Zhu et al. 2017) or unstructured texts (Ghazvininejad et al. 2018), for making improve the accuracy of proper noun translation especially medical domain.

Freshly, a variety of knowledge-based text generation corpora have been proposed (Dinan et al. 2018; Moghe et al. 2018; Zhou et al. 2018b) to fill the gap where previous datasets do not provide knowledge grounding of the text generation (Bahdanau et al. 2014; Sutskever et al. 2014; Vaswani et al. 2017). However,

© Springer Nature Singapore Pte Ltd. 2020
J. Li and A. Way (Eds.): CCMT 2020, CCIS 1328, pp. 37–54, 2020.
https://doi.org/10.1007/978-981-33-6162-1_4

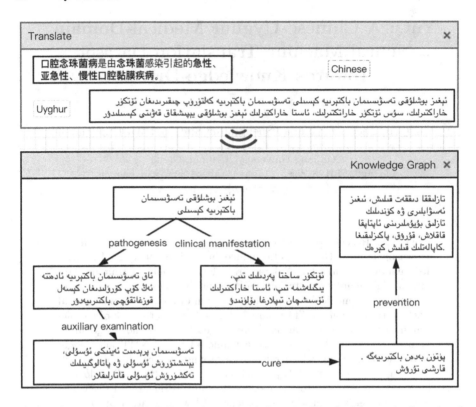

Fig. 1. An example in YuQ from the medical domain. The **bold** text is the related knowledge that is utilized in NMT.

these datasets are not suitable for the medical domain or knowledge planning through neural machine translation based on knowledge. OpenDialKG (Moon et al. 2019) and DuConv (Wu et al. 2019) use knowledge graphs as knowledge resources. However, for knowledge-grounded NMT datasets still have the gap.

In this paper, As given in Fig. 1, we propose YuQ, a Chinese-Uyghur neural machine translation dataset towards the medical domain, which is suitable for modeling knowledge interactions in machine translation in the medical domain, including knowledge planning, knowledge grounding, knowledge adaptations, etc. YuQ 65K utterances 130K parallel corpus in the medical domain. Each sentence is annotated with related knowledge entities in the knowledge graph, Its effect is as supervision for knowledge interaction modeling. Furthermore, YuQ contains medical topics, which manually annotated accurately with higher quality than other datasets. The relations of entity are explicitly defined in the knowledge graph. We provide a benchmark to evaluate both generation- and retrieval-based neural machine translation models on the YuQ dataset with/without access to the medical knowledge. Results show that knowledge-based contributes to the advancement of these models while existing models are still not strong enough to deliver knowledge-coherent NMT, indicating a large space for future work.

In summary, this paper makes the following contributions:

- We construct a new dataset, YuQ, for knowledge-driven neural machine translation in Chinese-Uyghur. YuQ 130K utterances in medical domains.
- YuQ provides a benchmark to evaluate the ability of neural machine translation with access to the corresponding knowledge in medical domains. The corpus can empower the research of not only knowledge-grounded machine translation text generation but also domain adaptation or transfer learning between similar domain or dissimilar domains.
- We provide benchmark models on this corpus to facilitate further research and conduct extensive experiments. Results show that the models can be enhanced by introducing background knowledge, but there is still much room for further research.

2   Related Work

Recently, neural machine translation has been largely advanced due to the increase of publicly available machine translation data (Bahdanau et al. 2014; Sutskever et al. 2014; Vaswani et al. 2017). However, the lack of annotation of background information or related knowledge results in a significant bottleneck in medical term translation, where the translation accuracy of medical terms needs to improve. These models produce a translation that is substantially different from those humans translate, which largely rely on background knowledge.

To facilitate the development of NMT models that mimic human translate, there have been several knowledge-grounded corpora proposed. (Duan et al. 2020) proposes a new NMT method that is based on no parallel sentences but can refer to a ground-truth bilingual dictionary. This new task can effectively improve the accuracy of the translation of specialized words in the medical domain. However, the Perplexity of translated sentences is not as well as Seq2Seq architecture. (Chen et al. 2020) considers the importance of the word in the sentence meaning and design a content word-aware NMT to improve translation performance. However, the accuracy of generated machine translation for medical terminology is often not controllable, resulting in some mistakes in the generated results. (Hao et al. 2019) presents multi-granularity self-attention (MG-SA): a neural network that combines multi-head self-attention and phrase modeling and can capture useful medical-domain phrase information at various levels of granularities. (Sokolov and Filimonov 2020) presents an automatic natural language generation system, capable of generating both human-like interactions and annotations by the means of paraphrasing to solve manual annotations are expensive and time-consuming.

To obtain the high-quality knowledge-grounded datasets, some studies construct from scratch with human annotators, based on the unstructured text or structured knowledge graphs. For instance, several datasets (Gopalakrishnan

et al. 2019; Zhou et al. 2018b, 2020) have human conversations where participants have access to the unstructured text of related background knowledge.
while OpenDialKG (Moon et al. 2019) and DuConv (Wu et al. 2019) build up
their corpora based on structured knowledge graphs. (Young et al. 2018) proposes to explicitly augment input text with concepts expanded via 1-hop relations where KG triples are represented in the sentence embeddings space. (He
et al. 2017) propose a system which iteratively updates KG embeddings and
attends over connected entities for response generation. However, several challenges remain to scale the simulated knowledge graph, for knowledge augmented
text generation, (Ghazvininejad et al. 2018; Long et al. 2017; Parthasarathi and
Pineau 2018) uses embedding vectors obtained from external knowledge sources,
Wikipedia, free-form text, etc. as an auxiliary input to the model in dialog generation. Knowledge graphs can provide rich structured knowledge facts for better
language understanding, (Zhang et al. 2019) utilize both large-scale textual corpora and KGs to train an enhanced language representation model (ERNIE),
which can take full advantage of lexical, syntactic, and knowledge information
simultaneously.

3   Datasets

The general method of constructing a parallel corpus is to collect, sort, mark, pre-
serve and utilize professional corpus software for parallel processing and retrieval
of the bilingual corpus. This paper is slightly different. In the processing of
Chinese corpus, automatic line partitioning is carried out first, and the text is
translated manually according to the line partition, which avoids the line labeling and alignment processing of the corpus. In the later retrieval, the method of
combining professional corpus software and self-built retrieval system is adopted.

3.1   Data Collection

By searching a huge number of literatures and investigating in the hospital, a
Chinese corpus from the general practitioner diagnosis and treatment system is
finally determined. The data collected covered seven clinical disciplines: internal medicine, surgery, pediatrics, obstetrics and Gynecology, infectious diseases,
dermatology, and Venereology, and five sense organs science. Each diagnosis and
treatment article was retrieved by using the word crawl tool text, and a storage directory is established according to the department name and disease type.
The disease name of a single diagnosis and treatment article was stored as a
TXT file name, and the storage format was UTF-8 A total of 593 articles, 7
department catalogs, 65 disease catalogs, and 593 disease diagnosis and treatment corpora have been built. The corpus data is from clinical diagnosis in the
hospital, and the content is authentic and representative. The balance of the
corpus is fully considered in the collection. The proportion of data collected
by each department is respectively Results: internal medicine 26.78%, surgery
15.17%, pediatrics 13.59%, obstetrics and Gynecology 10.92%, infectious diseases

12.09%, dermatology and Venereology 10.13%, facial science 11.30%, basically meet the actual needs of patients, and reflect the medical language style and characteristics.

3.2   Corpus Preprocessing

Chinese medical and health data were collected manually, totaling 45,216 sentences. The data cover 12 major clinical disciplines: infectious diseases, dermatology, and venereology, facial features, epidemiology, internal medicine, surgery, pediatrics, obstetrics and gynecology, neuropathy, psychiatry, ophthalmology, and stomatology, totaling 739 diseases. The collection contents for each disease include etiology and pathology; Diagnosis and differential diagnosis; Clinical manifestations; Inspection, auxiliary inspection, and laboratory inspection; Therapy and physical therapy; Prevention, etc. The acquisition of medical texts is a relatively difficult task, and its text preprocessing is also quite difficult. General data preprocessing methods are applied to medical texts, but the effect is not significant, and medical words are often scattered. For example, the word "da chang gan jun" is divided into two words "da chang" and "gan jun" in the preprocessing process. The obtained processing results cannot be directly used for translation model training. The input data set suitable for model training needs to be obtained through text garbled code filtering, length ratio filtering, text word segmentation, and other steps in advance. After denoising, the corpus is divided into three levels according to the UTF-8 format: root directory, Department directory and disease category directory.

3.3   Annotation

The actual work of translation processing is after the corpus is automatically entered into the database. At this time, the work of line segmentation and text entry into the database has been completed. Translators translate according to the prescribed format, avoiding the problem of alignment.

3.4   Knowledge Graph Construction

The sparsity and the huge scale of the knowledge are difficult to handle, the annotated medical corpus is expensive, and the knowledge of these medical entities contains both structured knowledge triples and unstructured knowledge texts, which make the task more general but challenging. After filtering the start entities which have few knowledge triples, the medical domain contains 215 start entities, respectively. After filtering the start entities, we built the knowledge graph. Given the start entities as seed, we build their neighbor entities within three hops. We merged the start entities and these build entities (nodes in the graph) and relations (edges in the graph) into a domain-specific knowledge graph for medical domains. The statistics of the knowledge graphs used in constructing YuQ are provided in Table 1 and Table 2.

Table 1. Statistics of the knowledge graph entity types of YuQ

Entity type	Explain	Number	Example
Test	Diagnostic Inspection Items	76	Blood sugar, urinary ketone body
Disease	Disease	23	Diabetic cardiomyopathy
Drug	Drug	73	Glibenclamide, repaglinide
Food	Food	19	Protein, fat
Symptom	Symptoms of disease	24	Drink more, eat more, urinate more
Total	Total	215	

Table 2. Statistics of the knowledge graph relationship types of YuQ

Relationship Type	Explain	Number	Example
Belongs_to	Belong to	2	<type 1 diabetes, belonging_to, diabetes>
Acompany_with	complicating disease	18	< diabetic cardiomyopathy, Acompany_with, diabetic microangiopathy >
Cure	Therapeutic	101	< metformin,cure,glyburide >
No_eat	Avoid food for diseases	10	<disease,No_eat,wine>
prevention	therapy method	9	<disease,prevention,sea fish>
Symptom	Disease symptoms	28	<disease,Symptom,urine>
auxiliary_examination	Check the diagnosis	287	<disease,auxiliary_examination,insulin>
Total	Total		465

4 Corpus Analysis

Chinese-Uyghur medical parallel corpus is a special corpus. By building a the-
saurus, analyzing the frequency of words, we can make an objective analysis of
the lexical features, determine the position and nature of different words in the
lexical list in the medical corpus, and reveal the distribution law of lexical frequency phenomenon. At the same time, we compare the self-built corpus with
other large-scale general corpora to further statistically analyze the importance
of different words in the special corpus.

4.1 Lexical Feature Analysis

4.1.1 Construct Vocabulary

Using the EmEditor tool to replace all part-of-speech tags in the segmented
corpus with spaces, A corpus separated by spaces is formed. According to the
decreasing order of the occurrence frequency of each word, i.e. High-frequency
words are ranked first and low-frequency words are ranked second, and the words
are numbered with natural numbers. The highest occurrence frequency is level
1, followed by level 2. Rank is used to represent the word-level sequence and
freq is used to represent the occurrence frequency of words in the corpus, thus
constructing the vocabulary shown in Table 3:

Table 3. Word frequency statistics. Word is a segmented word in the corpus. P is the probability that words appear in the corpus; Ln(r) and Ln(f) are used to calculate the logarithm of Rank and freq respectively.

Rank	Word	Freq	P	Ln(r)	Ln(f)
1	Treatment	2840	0.014196167	0	7.9515595
2	Onset	1488	0.007437992	0.6931472	7.305188
3	Symptoms	1428	0.0071380725	1.098612 3	7.26403
4	Occurrence	1162	0.005808431 7	1.386294 4	7.057898
5	Cause	1095	0.005473522	1.609 438	6.9985094
6	Patient	1033	0.005163606	1.7917595	6.9402223
7	General	957	0.0047837086	1.9459101	6.8638034
8	Serious	916	0.0045787636	2.0794415	6.8200164
9	Operation	866	0.004328831	2.1972246	6.763885
10	mg	832	0.004158877 3	2.3025851	6.7238326

4.1.2 Statistical Analysis of Word Frequency

Make statistics on the vocabulary, A total of 14,470 different words were acquired, Of these, 5,703 words appear only once, 39.41% of the total. Different from the general corpus, Most of the words with frequency 1 in this corpus are professional words in the medical field. Meaning. 121 words appear twice, 14.66% of that total. The word appearing more than three-time, 45.93% of the total. After analyzing the results of word frequency, That is, the 5% word appears only once, 20% word appears twice. But there is a slight gap, The main reason

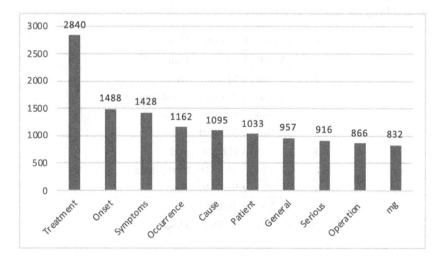

Fig. 2. The top ten high-frequency words and frequency of Chinese Medical Corpus.

is that there are many professional terms in the medical corpus, Word segmentation algorithm needs to be further improved, In addition, the corpus segmented by the current word segmentation algorithm, It also contains a large number of English strings, Chinese-English and English-Chinese mixed words, 323 with the frequency of 1, 104 with the frequency of 2 and 167 with other frequencies, which also conforms to the characteristics of medical corpus and includes a large number of transliterated and abbreviated foreign words, such as tc, r globulin, etc. The 10 words with the highest frequency in the vocabulary list and their frequencies are shown in Table 2, which fully reflect the medical characteristics of corpus data.

4.2   Contrastive Analysis of Lexical Features

The People's Daily has a corpus of nearly 2 million words in January, with a wide range of contents and a huge amount of data. Taking this as a reference corpus, the corpus retrieval software AntConc is used to compare and analyze the frequency of each word and word in the Chinese medical corpus word and word frequency list with the frequency of the word and word in the reference corpus to reflect the importance and particularity of the word and word to the medical corpus.

The list of word and word frequency includes information: Rank (word/word level), freq (frequency), word (word/word) and keyness (significant difference in frequency, the frequency difference between the same word and word in the two corpora, the greater the difference, the greater the keyness value). Some comparative statistical results are shown in Table 4.

Table 4. Statistical analysis for contrast of word frequency characteristics between our corpus and People's Daily corpora.

Rank	Freq	Keyness	Word
1	6 396	12213.213	Can
2	5 462	9 915.505	Or
3	2 840	6 381.669	Treatment
4	2 260	5 228.570	those
5	1 846	4 577.089	sex
6	2 661	3 756.796	and
7	1 428	3 519.548	Symptoms
8	1 493	2 949.540	Heart
9	1 707	2 469.261	as
10	1 144	2 386.947	Disease

4.2.1   Comparative Result Analysis

Corpus data using existing segmentation tools, based on Chinese medicine, shard 16637 words, which "Word" word frequency is greater than the reference corpus, 14050, 2587 less than the reference corpus, respectively constructed two-word frequency table.

Observe word frequency table of 14050 words, found keyness values by the maximum first became smaller, close to zero, until it is equal to zero, according to the keyness values change, the analysis of word frequency table is as follows:

In the first part, the value of keyness is very large at the beginning. Keyness >5 is taken as the boundary, and there are 7840 words, which are most commonly used in medical treatment, such as treatment and patient.

The second part, in order to 0 or less keyness 5 or less as the boundary, a total of 6210 words, at this time is divided into two cases: (1) to 0 or less keyness 5 or less and freq = 1, a total of 5089 words, observed that these words not only keyness value is small, gradually tends to zero, word frequency and minimum, these words are not commonly used for the two corpora, several medical field has the characteristics of medical is not commonly used words, such as early focal infarction disease, diffuse peritoneal infection, etc. With 321 as the letter combinations, such as athabasca, arvd. Athabasca, Arvd is acute obstructive suppurative cholangitis, respectively, the abbreviation of right ventricular cardiomyopathy arrhythmia caused by sex. (2) 0 or less keyness 5 or less and freq > 1, a total of 1121 words, this part of the vocabulary, freq value is very high, when keyness approach to find these words, such as, in the early morning, belong to the more commonly used words, basic is commonly used in the medical corpus, also commonly used in People's Daily corpus, frequency is similar in the two corpora.

Then Observe word frequency table of 2587 words:

(1) Keyness value is the largest at first and then decreases from the maximum. Contrary to the first part, when keyness value is large, it is all the data with high word frequency in the corpus of People's Daily, such as China, problems, development, etc.

(2) When keyness value is small and FREq = 1, the specificity of words cannot be seen, which is related to the fact that People's Daily is a general corpus. (3) when the keyness value smaller and larger freq, a total of 618 words, belong to two corpora are more frequently used vocabulary, but inadequate medical characteristics, such as a hospital bed, etc. After statistical analysis, combined with artificial proofreading, easily from 7840 words and 5089 words, sort out the medical special corpus theme vocabulary. Through the analysis of the above characteristics, not only reflects the corpus itself vocabulary characteristics, common vocabulary, vocabulary, etc. That validates whether corpus construction and late for further study of natural language processing technology to lay the foundation of medicine.

5 Experiments

5.1 Models

As provided baseline models for knowledge-driven NMT modeling, we evaluate such models on our corpus generation-based and retrieval-based models. To investigate the knowledge information annotation results, we evaluate the models with/without introducing to the knowledge graph of our dataset.

5.1.1 Generation-Based Models

Language Model (LM) (Bengio et al. 2003): We train a language model that maximizes the log likelihood: $logP(x) = \sum_t logP(x_t|x < t)$, where x denotes a long sentence that sequentially concatenates all the utterances of a machine translation.

Seq2Seq (Sutskever et al. 2014): An encoder-decoder model. The input of the encoder is the concatenation of the past $k-1$ utterances, while the target output of the decoder was the $k - th$ utterance. If there are fewer than $k - 1$ sentences in the NMT history, all the past utterances would be used as input.

RNNSearch (Bahdanau et al. 2014) RNNSearch is to improve the performance of Seq2Seq by the attention mechanism, where each word in Y is conditioned on different context vector c, with the observation that each word in Y may relate to different parts in x. In particular, y_i corresponds to a context vector c_i, and c_i is a weighted average of the encoder hidden states $h_1, ..., h_T$:

$$c_i = \sum_{j=1}^{T_x} a_{ij} h_j \tag{1}$$

where $a_{i,j}$ is computed by:

$$\alpha = \frac{\exp(e_{ij})}{\sum_{k=1}^{T} \exp(e_{ik})} \tag{2}$$

$$e_{ij} = g(s_{t-1}, h_j) \tag{3}$$

where g is a multilayer perceptron.

Transformer (Vaswani et al. 2017): Transformer abandons the recurrent network structure of RNN and models a piece of text entirely based on attention mechanisms. The most important module of the coding unit is the Self-Attention module, which can be described as:

$$Attention(Q, K, V) = Softmax\left(\frac{Q\mathbf{K}^{\mathrm{T}}}{\sqrt{d_k}}\right)V \tag{4}$$

To extend the ability of the model to focus on different locations and to increase the representation learning capacity of subspaces for attention units, Transformer adopts the "multi-head" mode that can be expressed as:

$$MultiHead(Q, K, V) = Concat(head_1, ..., head_h)W^O \qquad (5)$$

$$head_i = Attention(QW_i^Q, KW_i^K, VW_i^K) \qquad (6)$$

THUMT (Zhang et al. 2017): THUMT is an open-source toolkit for neural machine translation developed by the Natural Language Processing Group at Tsinghua University and a new implementation developed with TensorFlow.

5.1.2 Retrieval-Based Model

BERT (Devlin et al. 2019): We adapt this deep bidirectional transformers (Vaswani et al. 2017) as a retrieval-based model. For each utterance, we extract medical keywords and retrieve 10 translation candidates. The training task is to predict whether a candidate target utterance is the fitting source utterance given the source utterance where a sigmoid function is used to output the conditional probability $p(x_t|x_{0:t-1})$ can be modeled by a probability distribution over the vocabulary given linguistic context $x_{0:t-1}$. The context $x_{0:t-1}$ is modeled by neural encoder $f_{enc}(\cdot)$, and the conditional probability:

$$p(x_t|x_{0:t-1}) = g_{LM}\left(f_{enc}(x_{0:t-1})\right) \qquad (7)$$

where gLM(·) is the prediction layer. We select the candidate sentence with the largest probability.

5.1.3 Knowledge-Aware Models

A key-value memory module (Miller et al. 2016) is introduced to the aforementioned models to utilize the knowledge information. We treat all knowledge triples mentioned in an NMT as the knowledge information in the memory module. For a triple that is indexed by i, we represent the key memory and the value memory respectively as a key vector k_i and a value vector v_i, where k_i is the average word embeddings of the head entity and the relation, and v_i is those of the tail entity. We use a query vector q to attend to the key vectors $k_i(i = 1, 2, ...)$: $a_i = softmaxx_i(q^T k_i)$, then the weight sum of the value vectors $vi(i = 1, 2, ...), v = \sum_i a_i v_i$, is incorporated into the decoding process (for the generation-based models, concatenat with the initial state of the decoder) or the classification (for the retrieval-based model, concatenat with the <CLS> vector). For Seq2Seq, q is the final hidden state of the encoder. For RNNSearch and Transfomer, we treat the context vector as the query, while for BERT, the output vector of <CLS> is used.

Our dataset has a sentence-level annotation of the triples of knowledge used by each utterance. In order to compel the knowledge-aware models to attend to the KG triples, we applied an extra loss of focus.

$$L_att = -\frac{1}{|truth|} \sum_{i \in truth} \log a_i \qquad (8)$$

where truth is the set of indexes of triples that are used in the true response. The total loss are the weighted sum of $L^{(l)}$ and L_att:

$$L_{tot}^{(l)} = L_0^{(l)} + \lambda L_{att}, l \in g, r. \qquad (9)$$

The knowledge-enhanced BERT is initialized from the fine-tuned BERT, and the transformer parameters are frozen during training the knowledge related modules. The purpose is to exclude the impact of the deep transformers but only examine the potential effects introduced by the background knowledge.

5.2   Setup

We implement the above models with Pytorch while THUMT implement by tensorflow. The type of RNN network units is all GRU and the number of hidden units of GRU cells is all set to 200. ADAM as used to optimize all the models with the initial learning rate of 1×10^{-5} for BERT and 1×10^{-3} for others. The mini-batch sizes are set to 2 sentences for LM and 32 pairs of source- and target-sentence for Seq2Seq.

5.3   Automatic Evaluation

5.3.1   Metrics

We adopt BLEU, Rouge, and Perplexity as the evaluation metrics to measure the quality of the generated response. For BLEU, we employ the values of BLEU 1-4 and show the value of Rouge-1/2/L. Intuitively, the higher BLEU score and Rouge score mean more n-gram overlaps between the generated responses, and thereby indicate the better performance. Nevertheless, Perplexity is a well-established performance metric for generative text generation models. On the other hand, Perplexity explicitly measures the ability of the model to account for the syntactic structure of the dialogue, and the syntactic structure of each utterance and lower perplexity is indicative of a better model.

5.3.2   Results

The results are shown in Table 5. We analyze the results from the following viewpoints:

The Influence of Knowledge: In the medical domains, the knowledge-aware BERT model achieves the best performance in all of the metrics, as it retrieves the golden-truth response at a fairly high rate. The transformer-based models perform best in BLEU-k among all the generation-based baselines without considering the knowledge. Knowledge-aware Transformer has better results of

BLEU-k and better results of PPL, while the knowledge-enhanced Transformer–based models achieve the best metrics scores among all the generation-based models.

Comparison Between Models: In the medical domains, the knowledge-aware BERT model achieves the best performance in all of the metrics, as it retrieves the golden-truth response at a fairly high rate. The transformer-based models perform best in BLEU-k among all the generation-based baselines without considering the knowledge. Knowledge-aware Transformer has better results of BLEU-k and better results of PPL, while the knowledge-enhanced Transformer–based models achieve the best metrics scores among all the generation-based models.

Table 5. Automatic evaluation. The best results of generative models and retrieval models are in bold and underlined respectively. "+ know" means the models enhanced by the knowledge base.

Model	PPL	BLEU-1/2/3/4				Rouge-1/2/L		
LM	45.44	10.27	2.31	0.34	0.09	0.271	0.162	0.259
Seq2Seq	41.13	17.19	6.67	1.06	0.16	0.368	0.167	0.273
RNNSearch	40.45	20.97	8.40	1.71	1.27	0.387	0.124	0.248
Transformer	39.28	25.08	10.37	2.43	2.75	0.394	0.158	0.279
THUMT	21.91	24.22	12.40	2.71	2.27	0.384	0.207	0.313
BERT	37.32	27.63	14.32	3.35	3.13	0.427	0.216	0.314
Transformer+know	37.24	30.29	15.79	3.15	3.02	0.453	0.205	0.317
THUMT+know	37.91	30.41	18.43	3.72	3.01	0.498	0.237	0.349
BERT+know	33.11	33.14	20.54	4.93	3.91	0.592	0.481	0.591

5.4   Manual Evaluation

To better understand the quality of the generated responses from the semantic and knowledge perspective, we conducted the manual evaluation for knowledge-aware BERT, knowledge-aware RNNSearch, and Transformer, which have achieved advantageous performance in automatic evaluation.

5.5   Metrics

In terms of the fluency and coherence metrics, human annotators are asked to score a generated response.

Fluency (rating scale 0, 1, 2) is described as if the answer is normal and fluid:

- Grade 0 (bad): the grammatical mistakes are not articulate and challenging to comprehend.

- Grade 1 (fair): includes but yet clear grammatical errors.
- Grade 2 (good): humans generate it fluently and plausibly.

Coherence (rating scale is 0, 1, 2) is characterized as whether an answer to the context and knowledge information is valid and coherent:

- Grade 0 (bad): History is meaningless.
- Grade 1 (fair): important to the context, but not consistent with the details on expertise.
- Grade 2 (good): both context-relevant and consistent with background information.

5.6   Annotation Statistics

We randomly sampled about 500 contexts from the test sets and generated sentences by each model. These 1,500 parallel sentences pairs in total and related knowledge graphs are presented to three human annotators.

5.7   Results

The findings are seen in the Table 3. As can be shown, knowledge-aware BERT greatly outperforms other models in all dimensions in the medical realms, which correlates with automated evaluation performance. The Fluency is at 2.00 because all human-written sentences are the collected responses. The fluency scores of both generation-based models are approximately 2.00 suggesting that the translation produced is fluent and grammatical. The BERT and knowledge-aware BERT Coherence scores are higher than 1.00 but still have a big gap of 2.00, meaning that in most instances the translation produced is important to the background but

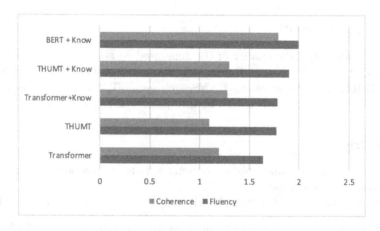

Fig. 3. Manual evaluation between three generative models. "+ know" means the models enhanced by knowledge information.

Table 6. Cases of the medical domain. Text is the knowledge used by the golden truth or the knowledge correctly utilized by the models.

Translation	Knowledge Triple		
	Head Entity	Relation	Tail Entity
CN: 眶下间隙蜂窝织炎	نېغىز بوشلۇقىنىڭ ئىڭەك يۈز قىسمى كۆزنەكسىمان توقۇلما ياللۇغى قىسمى كۆزنەكسىمان توقۇلما ياللۇغى	كېلىنىكىلىق نىپادىلىرى (clinical manifestation)	يۇقىرىقى لەۋ، نىشىشپىپ چىقىدۇ بۇرۇن-كالىپۇك نېرىقچىسى يۇقاپ كېپىدۇ
Ug: نېغىز بوشلۇقىنىڭ ئىڭەك يۈز قىسمى كۆزنەكسىمان توقۇلما ياللۇغى			
CN: 感染发生于眼眶下方，上颌骨前壁与面部表情肌之间。			
Ug: يۇقۇملىنىش قاپاقنىڭ ئاستى تەرىپى، ئۈستىنكى ئىڭەكنىڭ ئالدى ۋە يۈز قىسىملىرىدا كۆرۈنىدىغان كۆرۈنىش بولىدۇ.		داۋالاش (cure)	پۈتۈن بەدەن ئانتىبىئوتىكلىرىنى ماسلاشتۇرۇپ يۇقۇملىنىشقا قارشى تۇرۇش
CN:全身配合抗生素抗感染。			
Ug: پۈتۈن بەدەن ئانتىبىئوتىكلىرىنى ماسلاشتۇرۇپ يۇقۇملىنىشقا قارشى تۇرۇش			

not consistent with knowledge-aware facts. The Coherence score is substantially enhanced after integrating the knowledge information into BERT, as the knowledge information is more reflected in the produced translation.

5.8   Case Study

Some sample translations in the medical realms provided by knowledge-aware BERT are seen in Table 6. As we can see, knowledge-aware BERT is able to produce knowledge-based translation after the presentation of knowledge content, such as translation with expertise in the medical domain. However, it is still challenging for information-aware BERT to produce knowledge-coherent responses with respect to unstructured text awareness as modeling knowledge of unstructured texts requires more powerful models.

5.9   Ablation Study

To evaluate the contributions of key factors in our method, we perform an ablation study.

The influence of BPE on the Morphological segmentation of Uygur language. In order to verify the need for morphological segmentation of Uyghur language before using BPE technology, This paper compares the performance of BPE on Uyghur data without morphological segmentation and BPE on data after morphological segmentation under neural machine translation system respectively. According to Table 7, BLEU values of the same data set on the test set are 11.56 and 10.28 respectively. The former is 1.28 higher than the latter, and the improvement is not significant. Therefore, when BPE technology is adopted for the Uyghur language, morphological segmentation can be avoided, and BPE technology can effectively solve the problem of sparse data matrix.

Table 7. The influence of uyghur morphological segmentation on BLEU value.

Vocabulary	Uyghur morphological segmentation	BLEU
12000	+Morphological segmentation	11.56
12000	−Morphological segmentation	10.28

The Effect of Word List Size on Machine Translation Performance. The above experimental conclusions show that there is no great influence on whether Uyghur language is morphologically segmented and then BPE technology is used. The performance comparison experiment of neural machine translation methods based on the self-attention mechanism is continued under different vocabulary sizes. Table 8 experimental results show that Uyghur language does not undergo morphological segmentation, and the BLEU value with a vocabulary size of 32000 is 19.89 higher than that of Uyghur language with morphological segmentation and a vocabulary size of 12000. This shows that on the scarce resources and rich forms of Chinese-Uygur data set, Compared with morphological segmentation, The size of the word list can improve the performance of machine translation, The reason may be that morphological segmentation leads to the lack of semantic information at the word level, For enlarging the vocabulary, it can effectively reduce the number of unregistered words and save the effective information at the word level without losing. The neural network based on the self-attention mechanism can better learn the morphological structure and features of words, thus effectively improving the performance of machine translation.

Table 8. The influence of vocabulary size and morphological segmentation on neural machine translation.

Vocabulary	Uyghur morphological segmentation	BLEU
32000	−Morphological segmentation	30.17
12000	+Morphological segmentation	11.56

6   Conclusion and Future Work

In this paper, we propose a high-quality manually annotated Chinese-Uyghur medical-domain corpus for knowledge-driven neural machine translation, YuQ. It 130K utterances 65K parallel sentences, with an average length of 19.0. Each parallel sentence contains sentence-level annotations that map each utterance with the medical knowledge triples. The dataset provides a benchmark to evaluate the ability to model knowledge-driven translation. We provide generation and retrieval-based benchmark models to facilitate further research. Extensive experiments illustrate that NMT models can be enhanced by introducing knowledge, whereas there is still much room in knowledge-grounded neural machine translation modeling for future work. We hope that this dataset facilitates future research on the medical-domain neural machine translation problem.

Acknowledgments. We thank the anonymous reviewers for their valuable feedback. Qing Yu and Zhe Li are contributed equally to this research. This paper support by National Natural Science Foundation of China Research on the Construction of Chinese and Uygur Medical and Health Terms Resource Database Grant Number 61562082, and National Natural Science Foundation of China Research on Key Technologies of Uyghur-Chinese Phonetic Translation System Grant Number U1603262, Xinjiang Uygur Autonomous Region Graduate Research and Innovation Project Grant Number XJ2020G071, Dark Web Intelligence Analysis and User Identification Technology Grant Number 2017YFC0820702-3, and funded by National Engineering Laboratory for Public Safety Risk Perception and Control by Big Data (PSRPC).

References

Bahdanau, D., Cho, K., Bengio, Y.: Neural machine translation by jointly learning to align and translate. arXiv Computation and Language (2014)

Bengio, Y., Ducharme, R., Vincent, P., Jauvin, C.: A neural probabilistic language model. J. Mach. Learn. Res. **3**(Feb), 1137–1155 (2003)

Chen, K., Wang, R., Utiyama, M., Sumita, E.: Content word aware neural machine translation. In: Proceedings of the 58th Annual Meeting of the Association for Computational Linguistics, pp. 358–364 (2020)

Devlin, J., Chang, M.-W., Lee, K., Toutanova, K.: Bert: pre-training of deep bidirectional transformers for language understanding. In: NAACL 2019, pp. 4171–4186 (2019)

Dinan, E., Roller, S., Shuster, K., Fan, A., Auli, M., Weston, J.: Wizard of Wikipedia: knowledge-powered conversational agents. arXiv preprint arXiv:1811.01241 (2018)

Duan, X., et al.: Bilingual dictionary based neural machine translation without using parallel sentences. arXiv preprint arXiv:2007.02671 (2020)

Ghazvininejad, M., et al.: A knowledge-grounded neural conversation model. In: Thirty-Second AAAI Conference on Artificial Intelligence (2018)

Gopalakrishnan, K., et al.: Topical-chat: towards knowledge-grounded open-domain conversations. In: INTERSPEECH, pp. 1891–1895 (2019)

Hao, J., Wang, X., Shi, S., Zhang, J., Tu, Z.: Multi-granularity self-attention for neural machine translation. arXiv preprint arXiv:1909.02222 (2019)

He, H., Balakrishnan, A., Eric, M., Liang, P.: Learning symmetric collaborative dialogue agents with dynamic knowledge graph embeddings. In: Proceedings of the 55th Annual Meeting of the Association for Computational Linguistics (Volume 1: Long Papers), pp. 1766–1776 (2017)

Li, J., Galley, M., Brockett, C., Gao, J., Dolan, B.: A diversity-promoting objective function for neural conversation models. In: Proceedings of the 2016 Conference of the North American Chapter of the Association for Computational Linguistics: Human Language Technologies, pp. 110–119 (2016)

Long, Y., Wang, J., Xu, Z., Wang, Z., Wang, B., Wang, Z.: A knowledge enhanced generative conversational service agent. In: Proceedings of the 6th Dialog System Technology Challenges (DSTC6) Workshop (2017)

Miller, A., Fisch, A., Dodge, J., Karimi, A.H., Bordes, A., Weston, J.: Key-value memory networks for directly reading documents. In: Proceedings of the 2016 Conference on Empirical Methods in Natural Language Processing, pp. 1400–1409 (2016)

Moghe, N., Arora, S., Banerjee, S., Khapra, M.M.: Towards exploiting background knowledge for building conversation systems. In: Proceedings of the 2018 Conference on Empirical Methods in Natural Language Processing, pp. 2322–2332 (2018)

Moon, S., Shah, P., Kumar, A., Subba, R.: Opendialkg: explainable conversational reasoning with attention-based walks over knowledge graphs. In: Proceedings of the 57th Annual Meeting of the Association for Computational Linguistics, pp. 845–854 (2019)

Parthasarathi, P., Pineau, J.: Extending neural generative conversational model using external knowledge sources. In: Proceedings of the 2018 Conference on Empirical Methods in Natural Language Processing, pp. 690–695 (2018)

Shang, L., Lu, Z., Li, H.: Neural responding machine for short-text conversation. In: Proceedings of the 53rd Annual Meeting of the Association for Computational Linguistics and the 7th International Joint Conference on Natural Language Processing (Volume 1: Long Papers), pp. 1577–1586 (2015)

Shao, L., Gouws, S., Britz, D., Goldie, A., Strope, B., Kurzweil, R.: Generating long and diverse responses with neural conversation models (2016)

Sokolov, A., Filimonov, D.: Neural machine translation for paraphrase generation. arXiv preprint arXiv:2006.14223 (2020)

Sutskever, I., Vinyals, O., Le, Q.V.: Sequence to sequence learning with neural networks. In: Advances in Neural Information Processing Systems, pp. 3104–3112 (2014)

Vaswani, A., et al.: Attention is all you need, pp. 5998–6008 (2017)

Wen, T.-H., Gasic, M., Mrksic, N., Su, P.-H., Vandyke, D., Young, S.: Semantically conditioned LSTM-based natural language generation for spoken dialogue systems. In: Proceedings of the 2015 Conference on Empirical Methods in Natural Language Processing, pp. 1711–1721 (2015)

Wu, W., et al.: Proactive human-machine conversation with explicit conversation goal. In: Proceedings of the 57th Annual Meeting of the Association for Computational Linguistics, pp. 3794–3804 (2019)

Young, T., Cambria, E., Chaturvedi, I., Huang, M., Zhou, H., Biswas, S.: Augmenting end-to-end dialogue systems with commonsense knowledge. In: Thirty-Second AAAI Conference on Artificial Intelligence (2018)

Zhang, J., et al.: Thumt: an open source toolkit for neural machine translation. arXiv preprint arXiv:1706.06415 (2017)

Zhang, Z., Han, X., Liu, Z., Jiang, X., Sun, M., Liu, Q.: ERNIE: enhanced language representation with informative entities. In: Proceedings of the 57th Annual Meeting of the Association for Computational Linguistics, pp. 1441–1451 (2019)

Zhou, H., Huang, M., Zhu, X.: Context-aware natural language generation for spoken dialogue systems. In: Proceedings of COLING 2016, the 26th International Conference on Computational Linguistics: Technical Papers, pp. 2032–2041 (2016)

Zhou, H., Young, T., Huang, M., Zhao, H., Xu, J., Zhu, X.: Commonsense knowledge aware conversation generation with graph attention. In: IJCAI, pp. 4623–4629 (2018a)

Zhou, K., Prabhumoye, S., Black, A.W.: A dataset for document grounded conversations. In: Proceedings of the 2018 Conference on Empirical Methods in Natural Language Processing, pp. 708–713 (2018b)

Zhou, H., Zheng, C., Huang, K., Huang, M., Zhu, X.: KdConv: a Chinese multi-domain dialogue dataset towards multi-turn knowledge-driven conversation. arXiv preprint arXiv:2004.04100 (2020)

Zhu, W., Mo, K., Zhang, Y., Zhu, Z., Peng, X., Yang, Q.: Flexible end-to-end dialogue system for knowledge grounded conversation. arXiv, pp. arXiv-1709 (2017)

Quality Estimation for Machine Translation with Multi-granularity Interaction

Ke Tian[1,2(✉)] and Jiajun Zhang[1,2]

[1] National Laboratory of Pattern Recognition, Institute of Automation, CAS, Beijing, China
{ke.tian,jjzhang}@nlpr.ia.ac.cn
[2] School of Artificial Intelligence, University of Chinese Academy of Sciences, Beijing, China

Abstract. Quality estimation (QE) for machine translation is the task of evaluating the translation system quality without reference translations. By using the existing translation quality estimation methods, researchers mostly focus on how to extract better features but ignore the translation oriented interaction. In this paper, we propose a QE model for machine translation that integrates multi-granularity interaction on the word and sentence level. On sthe word level, each word of the target language sentence interacts with each word of the source language sentence and yields the similarity, and the L_∞ and entropy of the similarity distribution are employed as the word-level interaction score. On the sentence level, the similarity between the source and the target language translation is calculated to indicate the overall translation quality. Finally, we combine the word-level features and the sentence-level features with different weights. We perform thorough experiments with detailed studies and analyses on the English-German dataset in the WMT19 sentence-level QE task, demonstrating the effectiveness of our method.

Keywords: Quality estimation · Neural machine translation · Multi-granularity

1 Introduction

In recent years, neural machine translation (NMT) [1–4] makes great progress, and quality estimation of machine translation methods has also received much attention. Usually, evaluating system quality is to calculate the BLEU [5] when there are one or more reference translations available. In the model prediction or practical applications, it is costly to collect high-quality reference translations for each translation. Quality estimation of machine translation is the task of evaluating the translation system quality without reference translations. The prediction results can quickly measure the quality of the system translation. It plays

Supported by organization x.

an indispensable guiding role in post-translation editing and computer-aided translation. In the QE task, sentence-level QE is a popular research topic. Most sentence-level QE tasks predict a score which indicates how much effort is needed to post-edit translations to be acceptable results as measured by the Human-targeted Translation Edit Rate (HTER). In general, sentence-level QE is seen as a supervised regression task. In traditional feature-based QE approaches, which has 17 features that describe the translation quality, such as translation complexity indicators, fluency indicators, and adequacy indicators, it exploits a support vector regression algorithm to score the translation. With the rapid development of deep learning in natural language processing (NLP), many researchers have applied the neural network model to the QE task. With pre-trained language models showing excellent performance in natural language downstream tasks, multilingual pre-trained language models attract researchers' attention, such as Multilingual BERT [6], XLM [7].

Most researchers only focus on how to extract better features but ignore the translation oriented interaction. Although the word vectors fully interact in the neural network model, we believe that more translation characteristics should be added for cross-lingual tasks such as translation quality estimation. Either between word or sentence translation pairs, more translation oriented features can be tapped in.

In this paper, in order to solve the above problems, we propose a translation quality estimation method that incorporates multi-granularity interaction, making full use of the interactive information on the word and sentence level. And this method achieves good results in the WMT19 sentence-level QE task on the English-German dataset. On the word level, each word of the target language sentence interacts with each word of the source language sentence and yields the similarity, and the L_∞ and entropy of the similarity distribution are employed as the word-level interaction score. In terms of sentence level, we calculate the similarity between sentence vectors by cosine similarity. We specifically analyze that the similarity of translated word pairs can effectively measure translation quality.

2   Related Work

Traditional baseline model QuEst++ [8] extracted features based on handcrafted rules and used SVM regression to predict the score. With the great success of deep neural networks on many tasks in natural language processing(NLP), many researchers have applied the neural network model to the QE task. Shah et al. [9] combined neural features that include word embedding features and neural language model features with other features extracted by QuEst++. Kim et al. [10–12] proposed the Predictor-Estimator framework, within which predictor is product quality vectors by a bidirectional RNN encoder-decoder with attention mechanism, and estimator uses quality vectors to predict the score. Li et al. [13] combined the two-stage predictor-estimator framework to extract more abundant features through joint training. Fan et al. [14] proposed "Bilingual expert" model which uses transformer [15] architecture as feature extractor.

These models have achieved good results by using the powerful feature extraction ability of neural networks. In the past two years, the pre-training models such as EMLo [16], GPT [17], and BERT have developed rapidly and greatly improved the performance of downstream tasks in natural language processing. Lu et al. [18] proposed a feature extraction method based on the multi-language pre-training language model so that the source and target language sentence can interact more intensively, which is of great help to the cross-language task.

Kepler et al. [19] proposed the model that mainly integrates different sub-models, such as APE-BERT, PREDEST-BERT, and PREDEST-XLM, etc. From their experimental results, the key to the superior performance of their model depends on the PREDEST-XLM sub-model. Zhou et al. [20] used the translation model as a feature extraction module, and mainly improved the "Bilingual Expert" model with a SOURce-Conditional ELMo-style (SOURCE) strategy. Hou et al. [21] employed bi-directional translation knowledge and large-scale monolingual knowledge to the QE task. Kim et al. [22] proposed a "bilingual" BERT using multi-task learning for machine translation quality estimation.

In the above methods of QE, researchers mostly focus on using different model to extract better features, such as neural networks using recurrence, convolution and self-attention. But they ignored the translation oriented interaction. For the disadvantages of the above model, we will propose a QE model for machine translation that integrates multi-granularity interaction on the word and sentence level.

3 Methodology

In this section, Fig. 1 shows the model architecture. Following the recent trend in the NLP task exploiting large-scale language model pre-training for a series of different downstream tasks, we used multilingual BERT as feature extractors. The features fuse the translation oriented interaction on the word and sentence level and they are used to predict HTER score.

3.1 Model Architecture

The model consists of a feature extractor that produces contextual token representations, and an estimator that turns these representations into predictions for sentence-level scores. Although the multilingual pre-trained language model is well suited to handle cross-language tasks, it is still a single language followed by the same language is used as input for pre-training. In order to adjust the model and make it compatible with the input combination of the source and target language sentences, we adopt a cross-language joint encoding method that uses bilingual parallel corpus to pre-training multilingual BERT. This significantly improves the performance of the model.

As a multilingual model, source and target language sentences need to be input into the model together, so that the words in and between sentences can fully interact and get better vector representation. We combine the two sentences as input according to the template: [CLS] target [SEP] source [SEP], where [CLS] and [SEP] are special symbols from BERT, denoting the beginning of the sentence and sentence separators, respectively. The pre-training sub-task is to predict whether the second sentence is the translation of the first sentence.

Instead of just using contextual token representations to predict the score, we allow the contextual token representations of source and target language sentences to further interact. In other words, interaction is to explicitly model translation oriented features on the word and sentence level. The details of the interaction will be described in the next section. Finally, the word-level and sentence-level translation oriented features fuse with contextual token representation to predict scores.

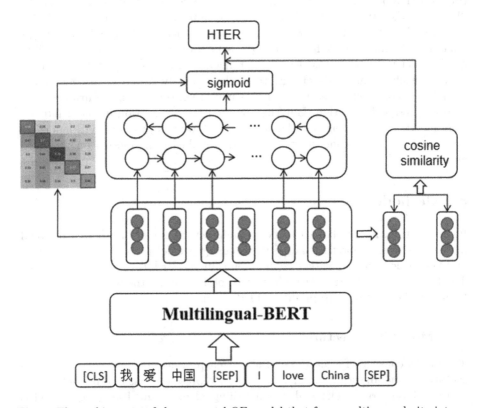

Fig. 1. The architecture of the proposed QE model that fuses multi-granularity interaction.

3.2   Multi-granularity Interaction

Word-Level Feature. Each word of the input bilingual sentences pair (source language sentence S, target language sentence T) is represented by pre-trained multilingual BERT $s = (s_1, s_2, ..., s_i)$ and $t = (t_1, t_2, ..., t_j)$. Compared to word embeddings, contextual embeddings provide different vector representations of the same word in different contexts. Since BERT uses subword encoding, we average the vectors of all subwords to represent a complete word. Next, we calculate the similarity matrix between each word in the target translation and each word in the source language sentence. The L_∞ and entropy of the similarity distribution are employed as that our method interaction core. Finally, the similarity scores of the translation word pairs are selected from the similarity matrix. There similarity scores are concatenated with entropy as features in the translation oriented interaction. As is shown in Fig. 2.

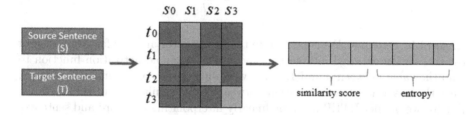

Fig. 2. An illustration of the word-level interaction.

we use the cosine metric to compute the similarity of each word pair between source and target language sentences. s_i denotes the ith token embedding of the source language sentence. t_j denotes the jth word embedding of the target language sentence.

$$sim_w = \frac{s_i \cdot t_j}{\|s_i\| * \|t_j\|} \tag{1}$$

The similarity distribution is the similarity score of a word in the target language sentence and each word in the source language sentence. The entropy of similarity distribution measures the confidence level of translation. H_j denotes entropy of the target language sentence's j^{th} word. p_i denotes probability after softmax.

$$H_j = -\sum_{i=1}^{n} p_i \cdot log p_i \tag{2}$$

Finally, the similarity score of word level and the entropy of similarity distribution are combined to form word-level interact feature. T denotes the number of words in the target language sentence.

$$E_i = Concat(sim_{w_1}, sim_{w_2}, \cdots, sim_{w_T}, H_1, H_2, \cdots, H_T) \tag{3}$$

Sentence-Level Feature. The vector representations of all the words in the target language sentence are averaged as the vector representation of the current sentence. And the calculation is shown in Formula 4. T denotes the number of words in the target language sentence. h_{tgt} denotes target language sentence vector. It the same for the source language sentence.

$$h_{tgt} = \frac{1}{T} \sum_{j=1}^{T} t_j \tag{4}$$

The similarity calculation of sentence level is same as word level. h_{src} denotes source language sentence vector. And the calculation is shown in Formula 5.

$$sim_s = \frac{h_{src} \cdot h_{tgt}}{\|h_{src}\| * \|h_{tgt}\|} \tag{5}$$

Ensemble Feature. We concatenate the word vector and the feature vector of the translation oriented interaction, and use the sigmoid activation function to map the value between 0 and 1. Then we subtract the cosine similarity from 1, since there is a negative correlation between cosine similarity and HTER value. Finally, we predict HTER score by linearly interpolating the word and sentence-level features.

$$hter = \lambda_1 \cdot sigmoid\left((E_w \bigoplus E_i) \cdot W\right) + \lambda_2 \cdot (1 - sim_s) \tag{6}$$

λ_1 and λ_2 denote word-level and sentence-level feature weights that are hyper-parameter. E_w denotes all word embedding. E_i denotes the translation oriented interaction feature embedding. sim_s denotes the similarity score of sentence vector between the source and target language sentence. $\bigoplus$ denotes vector concatenation operation. W denotes the learnable parameter matrix.

3.3 Model Training

Because the size of the training set for the QE task is too small to train the model, we use about 5 million bilingual parallel corpora to pre-train multilingual BERT. This also makes it more familiar with the input of the combination of source and target language sentences.

Assume that the training set for the QE task includes N source language sentences $x^{(n)}$, the target language sentences $y^{(n)}$, and the corresponding gold standard labels $HTER^{(n)}(n = 1, ..., N)$. The training objective is to minimize the mean square error over the training data:

$$R_{MSE} = \frac{1}{N} \sum_{i=1}^{n} (QE_{score}(x^{(n)}, y^{(n)}) - HTER^{(n)})^2 \tag{7}$$

4 Experiments

4.1 Dataset

The bilingual parallel corpus that we use for pre-trained multilingual BERT is officially released by the WMT17 Shared Task: Machine Translation of News1, including Europarl v7, Common Crawl corpus, News Commentary v12, and Rapid corpus of EU press releases. In the pre-training stage, we construct positive and negative samples from bilingual data. The positive sample is the parallel data correctly translated, while the negative sample is the source language sentence and the randomly sampled translation. The positive and negative samples are randomly shuffled to construct pre-train data.

In the QE experiment, to test the performance of the proposed QE model, we conduct experiments on the WMT19 sentence-level QE task for English-German (en-de) direction. The details of the dataset are shown in Table 1.

Table 1. Details of the en-de dataset of the WMT19 sentence-level QE task.

	Train	Dev	Test
Sentences	13442	1000	1023

4.2 Experimental Setup

In the experiment, we use the multilingual BERT after pre-training with bilingual parallel corpus, which has 12 Bi-transformer [13], and the total number of parameters is 1.1×10^8. During the training, we limit the number of training epoch as 3, learning rate 2×10^{-5}, batch size 32, max sequence length 128.

In the pre-training, we keep the default hyperparameter settings of the multilingual model. For the quality estimator module, the number of hidden units for forward and backward LSTM is 1000. We use a minibatch stochastic gradient descent algorithm and Adam to train the QE model.

4.3 Experimental Result

In this section, we will report the experimental results of our proposed model on the WMT19 sentence-level QE task for the English-German direction. And we list the results of other models in the WMT19 sentence-level QE task and the baseline model are listed in the Table 2.

Table 2. Results of the different models on the WMT19 sentence-level QE task.

System	Pearson	Spearman
UNBABEL	0.5718	0.6221
PREDEST-BERT	0.5190	–
CMULTIMLT	0.5474	0.5947
NJUNLP	0.5433	0.5694
ETRI	0.5260	0.5745
baseline	0.4001	0.4607
Our model	0.5496	0.5980

From the results in Table 2, we can see that our proposed model outperforms most of the baseline models. We also observe that our model underperforms the model UNBABEL. The reason is that UNBABEL is an ensembled model which integrates seven models. When we compare our model to their best single model PREDEST-BERT, we find that our model performs much better.

Then, we also compare the results of experiments that fuse different features, as shown in Table 3. We find that good performance can be achieved when all features are ensembled.

Table 3. Results of the models that fuse different features on the WMT19 sentence-level QE task.

System	Pearson	Spearman
BERT+LSTM	0.5057	0.5345
BERT+LSTM+word-level	0.5120	0.5606
BERT+LSTM+sentence-level	0.5332	0.5571
BERT+LSTM+sentence-leve+word-level	0.5496	0.5980

From the results in Table 3, both word-level features and sentence-level features are helpful to our tasks. It can be seen from the comparison that the features of sentence-level improve more based on the original model.

4.4   Word-Level Feature Analysis

We select an example from the dataset to explain the word level feature. As shown in Fig. 3 and Fig.4. It can be seen that 'dupliziert' is wrongly translated as 'Duiert'. According to the similarity matrix, the similarity score is slightly lower between the two words. However, other words are correctly translated, the corresponding similarity scores are much higher. As shown in Fig. 4, the similarity score corresponding to the punctuation is also low, and the reason is probably that the punctuation by itself does not take any meaning, unlike the nouns or verbs. Thus, we believe that the similarity score between words across languages can measure the quality of translation.

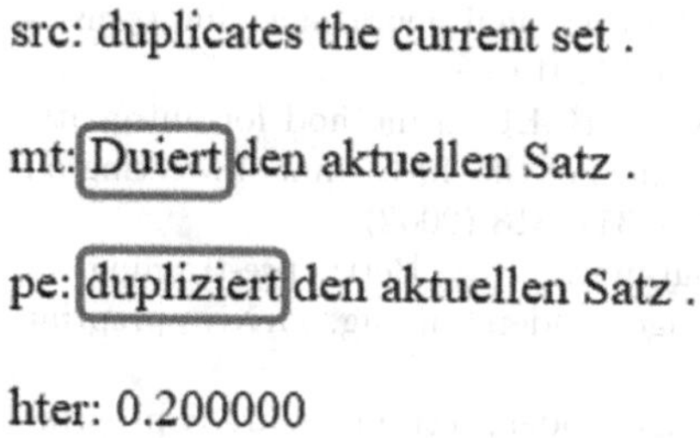

src: duplicates the current set .
mt: Duiert den aktuellen Satz .
pe: dupliziert den aktuellen Satz .
hter: 0.200000

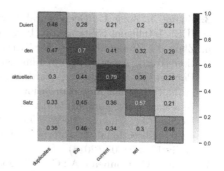

Fig. 3. An illustration of example.

Duiert 0.48 0.28 0.21 0.2 0.21
den 0.47 0.7 0.41 0.32 0.29
aktuellen 0.3 0.44 0.79 0.36 0.26
Satz 0.33 0.45 0.36 0.57 0.21
. 0.36 0.46 0.34 0.3 0.46
duplicates
the
current
set
.
1.0
0.8
0.6
0.4
0.2
0.0

Fig. 4. An illustration of word-level similarity matrix.

5 Conclusion

In this paper, we propose a translation quality evaluation method that fuses multi-granularity interaction. On the word level, each word of the target language sentence interacts with each word of the source language sentence, and the L_∞ and entropy of the similarity distribution are employed as the word-level interaction score. There similarity scores are concatenated with entropy as features in the translation oriented interaction. On the sentence level, the source and target language sentence vector similarity are used to measure the quality of the overall translation. Then, we predict the HTER score by linearly interpolating the word and sentence-level feature. Experimental results demonstrate that the method obtains more improvements over most models. And experimental results illustrate the validity of word-level interaction. In the future, we will conduct relevant experiments to verify the effect of the model in other language directions, and explore how to apply our approaches for word-level and document-level QE tasks.

Acknowledgments. The research work has been funded by the Natural Science Foundation of China under Grant No. U1836221 and 61673380. The research work in this paper has also been supported by Beijing Advanced Innovation Center for Language Resources and Beijing Academy of Artificial Intelligence (BAAI2019QN0504).

References

1. Bahdanau, D., Cho, K., Bengio, Y.: Neural machine translation by jointly learning to align and translate. arXiv preprint arXiv:1409.0473 (2014)
2. Zhang, J., Liu, S., Li, M., Zhou, M., Zong, C.: Bilingually-constrained phrase embeddings for machine translation. In: Proceedings of the 52nd Annual Meeting of the Association for Computational Linguistic, pp. 111–121 (2014)

3. Zhang, J., Zong, C.: Deep neural networks in machine translation: an overview. In: IEEE Intelligent Systems, pp. 16–25 (2015)

4. Zhou, L., Zhang, J., Zong, C.: Synchronous bidirectional neural machine translation. Trans. Assoc. Comput. Linguist. **7**, 91–105 (2019)

5. Papineni, K., Roukos, S., Ward, T., Zhu, W.-J.: BLEU: a method for automatic evaluation of machine translation. In: Proceedings of the 40th Annual Meeting on Association for Computational Linguistics, pp. 311–318 (2002)

6. Devlin, J., Chang, M.-W., Lee, K., Toutanova, K.: Bert: pre-training of deep bidirectional transformers for language understanding. arXiv preprint arXiv:1810.04805 (2018)

7. Lample, G., Conneau, A.: Cross-lingual language model pretraining. arXiv preprint arXiv:1901.07291 (2019)

8. Specia, L., Paetzold, G., Scarton, C.: Multi-level translation quality prediction with QuEst++. In: Proceedings of ACL-IJCNLP 2015 System Demonstrations, pp. 115–120 (2015)

9. Shah, K., Ng, R.W.M., Bougares, F., Specia, L.: Investigating continuous space language models for machine translation quality estimation. In: Proceedings of the 2015 Conference on Empirical Methods in Natural Language Processing, pp. 1073–1078 (2015)

10. Kim, H., Jung, H.-Y., Kwon, H., Lee, J.-H., Na, S.-H.: Predictor-estimator: neural quality estimation based on target word prediction for machine translation. ACM Trans. Asian Low-Resource Lang. Inf. Process. (TALLIP) **17**, 1–22 (2017)

11. Kim, H., Lee, J.-H.: Recurrent neural network based translation quality estimation. In: Proceedings of the First Conference on Machine Translation: Volume 2, Shared Task Papers, pp. 787–792 (2016)

12. Kim, H., Lee, J.-H., Na, S.-H.: Predictor-estimator using multilevel task learning with stack propagation for neural quality estimation. In: Proceedings of the Second Conference on Machine Translation, pp. 562–568 (2017)

13. Li, M., Xiang, Q., Chen, Z., Wang, M.: A unified neural network for quality estimation of machine translation. IEICE Trans. Inf. Syst. **101**, 2417–2421 (2018)

14. Fan, K., Wang, J., Li, B., Zhou, F., Chen, B., Si, L.: "Bilingual expert" can find translation errors. In: Proceedings of the AAAI Conference on Artificial Intelligence, pp. 6367–6374 (2019)

15. Vaswani, A., et al.: Attention is all you need. In: Advances in Neural Information Processing Systems, pp. 5998–6008 (2017)

16. Peters, M.E., et al.: Deep contextualized word representations. arXiv preprint arXiv:1802.05365 (2018)

17. Radford, A., Narasimhan, K., Salimans, T., Sutskever, I.: Improving language understanding by generative pre-training (2018). https://s3-us-west-2. amazonaws.com/openai-assets/researchcovers/languageunsupervised/language_ understanding_paper.pdf

18. Lu, J., Zhang, J.: Quality estimation based on multilingual pre-trained language model. J. Xiamen Univ. Nat. Sci. **59**(2) (2020). (in Chinese)

19. Kepler, F., et al.: Unbabel's participation in the WMT19 translation quality estimation shared task. In: Proceedings of the Fourth Conference on Machine Translation (Volume 3: Shared Task Papers, Day 2), pp. 78–84 (2019)

20. Zhou, J., Zhang, Z., Hu, Z.: SOURCE: SOURce-conditional elmo-style model for machine translation quality estimation. In: Proceedings of the Fourth Conference on Machine Translation (Volume 3: Shared Task Papers, Day 2), pp. 106–111 (2019)

21. Hou, Q., Huang, S., Ning, T., Dai, X., Chen, J.: NJU submissions for the WMT19 quality estimation shared task. In: Proceedings of the Fourth Conference on Machine Translation (Volume 3: Shared Task Papers, Day 2), pp. 95–100 (2019)
22. Kim, H., Lim, J.-H., Kim, H.-K., Na, S.-H.: QE BERT: bilingual BERT using multi-task learning for neural quality estimation. In: Proceedings of the Fourth Conference on Machine Translation (Volume 3: Shared Task Papers, Day 2), pp. 85–89 (2019)

Transformer-Based Unified Neural Network for Quality Estimation and Transformer-Based Re-decoding Model for Machine Translation

Cong Chen, Qinqin Zong, Qi Luo, Bailian Qiu, and Maoxi Li[✉]

Jiangxi Normal University, Nanchang, Jiangxi, China
chencong.jxnu@gmail.com, {zongqinqin,luoqi,mosesli}@jxnu.edu.cn,
qiubl@ecjtu.edu.cn

Abstract. In this paper, we describe our submitted system for CCMT 2020 sentence-level quality estimation subtasks and machine translation subtasks. We propose two models: (i) a Transformer-based unified neural network for the quality estimation of machine translation, which consists of a Transformer bottleneck layer and a bidirectional long short-term memory network that are jointly optimized and fine-tuned for quality estimation, and (ii) a Transformer-based re-decoding model for machine translation, which takes the generated machine translation outputs as the approximate contextual environment of the target language and synchronously re-decodes each token in the machine translation outputs. Experimental results on the development set show that the proposed approaches outperform the baseline systems.

Keywords: Machine translation · Quality estimation of machine translation · Re-decoding · Encoder-decoder model

1 Introduction

The 16th China Conference on Machine Translation (CCMT 2020) was organized around machine translation [10] evaluation tasks, which consist of four subtasks: bilingual translation, multilingual translation, speech translation, and the quality estimation of machine translation. The team of Jiangxi Normal University participated in two subtasks in the conference: the sentence-level quality estimation of machine translation and machine translation. The systems and related technologies we used for these two evaluation subtasks and the system performance for the development set are presented in this paper.

2 Model

2.1 Transformer-Based Unified Neural Network for the Quality Estimation of Machine Translation

The quality estimation of machine translation output is performed without relying on reference translations. Quality estimation plays an important role in

© Springer Nature Singapore Pte Ltd. 2020
J. Li and A. Way (Eds.): CCMT 2020, CCIS 1328, pp. 66–75, 2020.
https://doi.org/10.1007/978-981-33-6162-1_6

post-editing [6]. Sentence-level translation quality estimation is generally regarded as a regression problem. Features are extracted from source sentences and their machine translation outputs [1], and then input into a regression model to obtain a sentence quality score for the machine translation [4].

A bottleneck layer is generally defined as a multilayer neural network that abstracts the raw instance into a high-dimensional embedding in the deep neural network. The bottleneck layer and its output embeddings play an important role in transfer learning [9]. To fully use the bilingual associative knowledge learned from the bilingual parallel corpus through the Transformer model, we propose a Transformer-based unified neural network for quality estimation (TUNQE) model, which is a combination of the bottleneck layer of the Transformer model with a bidirectional long short-term memory network (Bi-LSTM), as shown in Fig. 1. The process by which the translation outputs to be estimated and the corresponding source sentences reach the top of the bottleneck layer through the trained unified neural network can be regarded as a feature extraction process for words in the machine translations. The Bi-LSTM layer converts word-level features into sentence-level features, which are input to a feed-forward neural network to predict the translation quality scores.

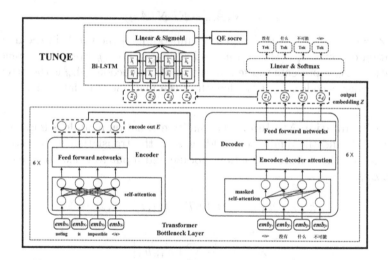

Fig. 1. TUNQE model architecture.

Feature Extraction Module. The feature extraction module is located in the bottleneck layer of the Transformer model (the bottom half of Fig. 1). This module is used to extract the quality embedding of the words in the machine translations to be estimated, namely, the word embedding Z of the output of the Transformer bottleneck layer.

The first step in extracting the quality embedding Z of the words in the translations is to encode the input source sentences X to yield the encoder output E:

$$A_e = LN\left(W_{emb}X + Attention\left(W_{emb}X, W_{emb}X, W_{emb}X\right)\right) \qquad (1)$$

$$E = LN\left(A_e + FFN\left(A_e\right)\right) \qquad (2)$$

where $Attention()$ is the self-attention function in the Transformer, $LN()$ is the layer normalization function in the Transformer, $FFN()$ is the position feed-forward neural network function [8], the symbol A_e represents the output embedding of the encoder's self-attention, and W_{emb} is the word embedding matrix.

The source sentences are encoded to obtain the representation E, which is input into the decoder with the machine translation to be estimated Y, and the quality embedding Z of the machine translations is extracted:

$$A_d = LN\left(W_{emb}Y + Attention\left(W_{emb}Y, W_{emb}Y, W_{emb}Y\right)\right) \qquad (3)$$

$$A_{ed} = LN\left(A_d + Attention\left(A_d, E, E\right)\right) \qquad (4)$$

$$Z = LN\left(A_{ed} + FFN\left(A_{ed}\right)\right) \qquad (5)$$

where $Z = (z_1, z_2, ..., z_{L_y})$ is the quality embedding of the words in the machine translation outputs. z_i is the quality embedding of the i_{th} word in the machine translations. A_d represents the self-attention of the encoder. A_{ed} is the attention of the encoder-decoder. L_y is the length of the machine translation Y.

Quality Estimation Module. The quality estimation module is implemented by Bi-LSTM, and the quality embedding Z of the translation to be estimated is obtained by the feature extraction module and input to calculate the quality score QE_{sorre}:

$$\overrightarrow{h}_{1:L_y}; \overleftarrow{h}_{1:L_y} = BiLSTM\left(z_1, z_2, \ldots, z_{L_y}\right) \qquad (6)$$

$$Z_{sen} = \frac{1}{L_y}\sum_{i=1}^{L_y}\left[\overrightarrow{h}_i; \overleftarrow{h}_i\right] \qquad (7)$$

$$QE_{sorre} = sigmoid\left(W_{qe}Z_{sen}\right) \qquad (8)$$

where the symbol $\overrightarrow{h}_i$ represents the hidden state of the i_{th} forward time-step of Bi-LSTM, and $\overleftarrow{h}_i$ represents the hidden state of the i_{th} backward time-step of Bi-LSTM. Z_{sen} is the sentence-level quality embedding of the machine translations, and W_{qe} is the weight parameters of the full connection layer in the quality estimation module.

2.2   Study of Re-decoding-Based Neural Machine Translation

In recent years, the Transformer [8], which exploited the self-attention mechanism in the encoder and in the decoder, significantly improved translation quality. However, the model usually generates a sequence token-by-token from left to right; hence, this autoregressive decoding procedure lacks the guidance of a future context, which is crucial to prevent undertranslation. To alleviate this issue, we propose a TransRedecoder model (Fig. 2), which employs a Mask-CURRENT attention matrix (Fig. 3(b)) to predict the re-decoding output sequence.

As shown in Fig. 2, the same encoder structure is used in the TransRedecoder model as in the Transformer model. The decoder of the TransRedecoder model is an identical layer. Unlike the masked matrix used in the Transformer decoder (Fig. 3(a)), the TransRedecoder model decoder employs the Mask-CURRENT attention matrix to fully use the machine translation generated by the Transformer as an approximate contextual environment of the target language. During the re-decoding, we enter the source language (src) and machine translation (mt) generated by the Transformer into the encoder and decoder, respectively. The former contents and the post contents of position j are combined in the machine translation of the target language generated by the Transformer to generate the re-decoding machine translation.

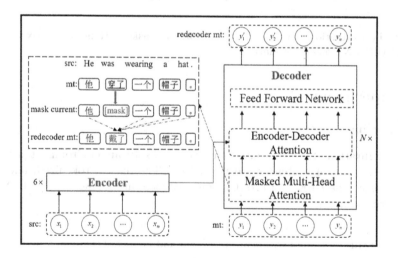

Fig. 2. TransRedecoder model architecture.

Given an input sentence $x = (x_1, x_2, \ldots, x_m)$ and its translation outputs $y = (y_1, y_2, \ldots, y_n)$, the model successively re-decodes each token in the translation outputs and generates a new machine translation output y'. As a result, every token in the re-decoded sequence y' fully uses the contextual information:

$$P\left(y' \mid x, y; \theta'\right) = \prod_{i=1}^{n} P\left(y_i' \mid x, y; \theta'\right) \tag{9}$$

where θ' represents the parameters of the TransRedecoder model.

As shown in Fig. 3(b), the attention matrix is utilized to combine the former contents and the post contents of position j in the machine translation of the target language generated by the Transformer to generate the re-decoding machine translation. When $t = 2$, the decoder applies the masked vector $(1, 1, 0, 1, 1)$ for masking "go", utilizes the contextual information "$\langle/s\rangle$, we, hiking, yesterday" of "go", and modifies "go" into "went". This helps solve the under-translation problem caused by the absence of the future text.

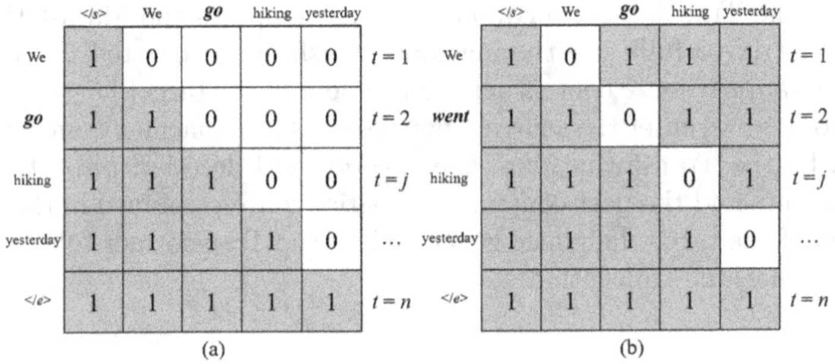

(a)

	$\langle/s\rangle$	We	**go**	hiking	yesterday	
We	1	0	0	0	0	$t = 1$
go	1	1	0	0	0	$t = 2$
hiking	1	1	1	0	0	$t = j$
yesterday	1	1	1	1	0	...
$\langle/e\rangle$	1	1	1	1	1	$t = n$

(b)

	$\langle/s\rangle$	We	**go**	hiking	yesterday	
We	1	0	1	1	1	$t = 1$
went	1	1	0	1	1	$t = 2$
hiking	1	1	1	0	1	$t = j$
yesterday	1	1	1	1	0	...
$\langle/e\rangle$	1	1	1	1	1	$t = n$

Fig. 3. Two attention matrices: (a) Transformer masked attention matrix and (b) Mask-CURRENT attention matrix.

3   Experiment

3.1   Setting

The configuration of the computer hardware and software environment is shown in Table 1. English sentences are normalized, lowercased, tokenized, and segmented using the BPE subword. Chinese sentences are segmented by the Stanford word segmenter.

Table 1. Computer operating system and hardware configuration.

Operating system	CPU	Memory	GPU
Ubuntu19.04 LTS	Intel i5-6500	32G	GeForce GTX 2080Ti

To evaluate quality estimation, we pre-train the Transformer model using the data provided by the CWMT2018 news translation task. The TUNQE model is jointly optimized and fine-tuned with the training set provided by the CCMT2020 quality estimation tasks. Statistics for the corpus size are shown in Table 2.

Table 2. Statistics for translation quality estimation evaluation corpus.

	Direction	Training set	Development set	Test set
CWMT2018	Parallel corpus	6M	3000	3000
CCMT2020	zh-en	10070	1143	4211
	en-zh	14789	1381	4355

The training set used to evaluate the machine translation subtask is entirely provided by the CCMT2020 machine translation subtask. Statistics for the corpus size are shown in Table 3.

Table 3. Statistics for translation quality estimation evaluation corpus.

	Pair of sentences			Number of tokens		
	Training set	Development set	Test set	Training set	Development set	Test set
en-zh	9M	10K	1997	1375M	2.5M	0.2M
zh-en	9M	10K	2000	1375M	2.5M	0.2M

The TUNQE model is developed based on the Facebook fairseq open source toolkit [5]. The Transformer bottleneck layer is pre-trained by using the Adam optimizer $(\beta_1 = 0.9, \beta_2 = 0.98, \varepsilon = 10^{-9})$, where the learning rate lr = 0.0007 and the minimum learning rate min_lr = 10^{-9}. The SGD optimizer is adopted when the Transformer bottleneck layer and the Bi-LSTM layer are jointly optimized and fine-tuned with a quality estimation training set. The learning rate is fixed at 0.05.

The parameters of the Transformer translation model are consistent with the Transformer-base proposed by Vaswani [8]. The decoder of the TransRedecoder model is an identical layer, and the remaining model parameters are consistent with those of the Transformer model. The feed forward neural network layer has a dimensionality of 2048. We employ 8 parallel attention layers or heads. The Adam optimizer is used to train the model at a learning rate lr = 0.0003 and a minimum learning rate min_lr = 10^{-9}. To facilitate these residual connections, all the sublayers in the model, as well as the embedding layers, produce outputs of dimension 512.

3.2   Results

Sentence-Level Quality Estimation Task. The performance of quality estimation is evaluated in terms of the Pearson correlation coefficient between the quality estimation and human judgments, and the Spearman correlation coefficient is used to measure the correlation between the rankings of the translation quality and human judgments. The higher the Pearson or Spearman correlation coefficient is, the higher the model performance is. The TUNQE model is tested on the CCMT2020 development set for sentence-level quality estimation. The experimental results are shown in Table 4.

Table 4. TUNQE results for CCMT2020 sentence-level QE dev set.

Model	Parallel corpus	en-zh		zh-en	
		Pearon	Spearman	Pearson	Spearman
TUNQE$_{SEP}$	CWMT 6 M	0.4476	0.3128	0.4877	0.4277
TUNQE		0.5055	0.3555	**0.5888**	0.4806
TUNQE$_{BERT}$		**0.5322**	0.3785	0.5735	0.4721

We assess the advantages of jointly training the unified neural network by comparing the performances of TUNQE and TUNQE$_{SEP}$. TUNQE$_{SEP}$ is a method of separately training the Transformer bottleneck and the Bi-LSTM layers using a bilingual parallel corpus and a sentence-level quality estimation training set, respectively. The experimental results in Table 4 show that the TUNQE method outperforms the TUNQE$_{SEP}$ method. The Pearson correlation coefficients of TUNQE are improved by 12.9% and 20.7% in the en-zh and zh-en directions, respectively, over those of TUNQE$_{SEP}$.

Li et al. verified that the integration of BERT contextual word embedding [2] can improve translation quality estimation by using the fluency features of the translation [11]. We apply this method to estimate the translation quality, where by BERT pre-trained word embedding in the translation is fused with the embedding extracted by TUNQE after the average pooling of the last 4 layers of representation, which is named TUNQE$_{BERT}$. The experimental results show that TUNQE$_{BERT}$ exhibits higher system performance than TUNQE.

Machine Translation Task. The final machine translation results of the CCMT 2020 en-zh and en-zh direction development sets are shown in Table 5. In the en-zh direction, the BLEU score of the re-decoding machine translation increases by 1.26, and the NIST score of the re-decoding machine translation increases by 0.15.

We verify the validity of the TransRedecoder model by using the same data post-processing and scoring approaches for the experimental results submitted by KSAI [3] and Baidu [7] for WMT2019 and utilize the TransRedecoder model to generate a re-decoding machine translation based on the original machine translation. KSAI used the Transformer [8] as a baseline system, trained the model with 24.22 M pairs of sentences, and then used data filtering, fine-tuning, back translation, model enhancement, model integration, and reordering techniques to improve the translation quality. Baidu used the big Transformer [8] as a baseline system. Baidu trained the model with 15.7 M pairs of sentences in the en-zh and zh-en directions, and back translation, joint training, knowledge distillation, fine-tuning, model integration and reordering technology were also used to improve the machine translation quality.

Table 5. Results of original and re-decoding machine translation for different machine translations of CCMT2020 dev sets. MT_O means the original machine translation and MT_R means the re-decoding machine translation.

| | Transformer | | | | KSAI | | | | Baidu | | | |
| | en-zh | | zh-en | | en-zh | | zh-en | | en-zh | | zh-en | |
	BLEU	NIST	BLEU	NIST	BLEU	NIST	BLEU	NIST	BLEU	NIST	BLEU	NIST
MT_O	31.52	7.84	25.15	7.02	42.42	9.14	40.25	9.09	42.49	9.22	40.95	9.21
MT_R	32.78	7.99	26.51	7.18	42.61	9.15	40.79	9.13	42.65	9.24	41.45	9.25
Δ	1.26	0.15	1.36	0.16	0.19	0.01	0.54	0.04	0.16	0.02	0.5	0.04

The re-decoding experimental results are shown in Table 5. In the en-zh direction, the TransRedecoder model increases the BLEU score by 0.19 and 0.16, respectively, and the NIST score by 0.01 and 0.02, respectively. In the zh-en direction, the TransRedecoder model significantly improves the BLEU scores by 0.54 and 0.50, respectively, and both improve the NIST scores by 0.04. Although the Baidu/ KSAI submitted systems achieved the best translation performance for the WMT19 test sets, the results in Table 5 show there is room for improvement. The novel re-decoding-based neural machine translation model, TransRedecoder, improves upon the quality of the machine translation.

3.3 Analysis

The re-decoding-based neural machine translation method is validated by the results in Table 6, which demonstrate an example of the original machine translation and re-decoding machine translation generated by the TransRedecoder model for the en-zh and zh-en directional dev sets of the CCMT2020 machine translation evaluation subtask. A comparison with the human reference translation clearly shows that the TransRedecoder model can effectively correct inaccurate target words in the machine translation. In the en-zh direction, the future context of "禁止" is utilized to generate the re-decoding words "拒绝", which is better with "提供 庇护" than "拒绝"; in the zh-en direction; "hope" is re-decoded by combining the future context of "meeting", and the re-decoded output "intension" is a better translation of the source word "会谈". The results demonstrate that the proposed TransRedecoder model can effectively utilize contextual information from the original machine translation to improve the quality of the re-decoding machine translation of the target language.

Table 6. Original machine translation and re-decoding machine translation.

Source	eighteen states and the district of columbia are supporting a legal challenge to a new u.s. policy that denies asylum to victims fleeing gang or domestic violence .
Reference	美国 18 个 州 和 哥伦比亚 特区 支持 对 一 项 新 政策 发起 法律 挑战 ， 这 项 政策 拒绝 向 逃离 帮派 或 家庭 暴力 的 受害者 提供 庇护 。
MT	十八 个 州 和 哥伦比亚 特区 正 支持 对 美国 的 一 项 新 政策 的 法律 挑战 ， 这 项 政策 禁止 为 逃避 帮派 或 家庭 暴力 的 受害者 提供 庇护 。
Re-decoding MT	18 个 州 和 哥伦比亚 特区 政府 支持 对 美国 的 一 项 新 政策 的 法律 挑战 ， 这 项 政策 拒绝 向 逃离 帮派 或 家庭 暴力 的 受害者 提供 庇护 。
Source	同时 他 也 表达 了 希望 与 玉城 会谈 的 意向 。
Reference	and he also expressed his intention to talk with tamaki .
MT	at the same time , he also expressed the hope of meeting with yucheng .
Re-decoding MT	at the same time , he also expressed his intention of talks with yucheng .

4   Conclusions

This paper is a description of a technical report from Jiangxi Normal University on CCMT2020 sentence-level translation quality estimation subtasks and machine translation evaluation subtasks. We propose a simple and effective unified neural network model based on the Transformer model to effectively improve the performance of sentence-level translation quality estimation, as well as propose a TransRedecoder model that employs the Mask-CURRENT attention matrix to use the context of the original machine translation to increase the quality of the machine translation.

Acknowledgements. This research has been funded by the Natural Science Foundation of China under Grant No.61662031 and 61462044. The authors would like to extend their sincere thanks to the anonymous reviewers who provided valuable comments.

References

1. Chen, Z., et al.: Improving machine translation quality estimation with neural network features. In: Proceedings of the WMT, pp. 551–555 (2017)
2. Devlin, J., Chang, M.W., Lee, K., Toutanova, K.: Bert: pre-training of deep bidirectional transformers for language understanding. In: Proceedings of the NAACL, pp. 4171–4186 (2018)
3. Guo, X., et al.: Kingsoft's neural machine translation system for wmt19. In: Proceeding of the ACL, pp. 196–202 (2019)

4. Li, M., Xiang, Q., Chen, Z., Wang, M.: A unified neural network for quality estimation of machine translation. IEICE Trans. Inf. Syst. **101**(9), 2417–2421 (2018)
5. Ott, M., et al.: Fairseq: a fast, extensible toolkit for sequence modeling. In: Proceedings of the NAACL, pp. 48–53 (2019)
6. Specia, L., Shah, K., De Souza, J.G., Cohn, T.: Quest-a translation quality estimation framework. In: Proceedings of the ACL, pp. 79–84 (2013)
7. Sun, M., Jiang, B., Xiong, H., He, Z., Wu, H., Wang, H.: Baidu neural machine translation systems for wmt19. In: Proceeding of the ACL, pp. 374–381 (2019)
8. Vaswani, A., et al.: Attention is all you need. In: Proceedings of the NIPS, pp. 5998–6008 (2017)
9. Yu, D., Seltzer, M.L.: Improved bottleneck features using pretrained deep neural networks. In: Proceedings of the INTERSPEECH, pp. 237–240 (2011)
10. 李亚超, 熊德意, 张民: 神经机器翻译综述. 计算机学报 **41**(12), 100–121 (2018)
11. 李培芸, 李茂西, 裘白莲, 王明文: 融合 bert 语境词向量的译文质量估计方法研究. 中文信息学报 **34**(3), 56–63 (2020)

NJUNLP's Machine Translation System for CCMT-2020 Uighur → Chinese Translation Task

Dongqi Wang, Zihan Liu, Qingnan Jiang, Zewei Sun, Shujian Huang[✉],
and Jiajun Chen

National Key Laboratory for Novel Software Technology, Nanjing University,
Nanjing, China
{wangdq,liuzh,jiangqn,sunzw}@smail.nju.edu.cn
{huangsj,chenjj}@nju.edu.cn

Abstract. This paper describes our submitted systems for CCMT-2020 shared translation tasks. We build our neural machine translation system based on Google's Transformer architecture. We also employ some effective techniques such as back translation, data selection, ensemble translation, fine-tuning and reranking to improve our system.

Keywords: Neural machine translation · Back translation · Fine-tuning · Ensemble translation · Reranking

1 Introduction

Neural networks have shown their superiority on machine translation [1,18] and other natural language processing tasks [5]. Self-attention based Transformer [19] has been the dominant architecture for neural machine translation. This paper describes our submission for CCMT-2020 Uighur → Chinese translation task.

We build our system based on Transformer [19] due to its superior performance and parallelism. Several techniques which have been proved effective are employed to boost the performance of our system.

We apply Byte Pair Encoding (BPE) [15] to reduce the sizes of vocabularies and achieve open-vocabulary translation. Tagged back-translation with top-k sampling [2,7,14] is used to improve translation performance with monolingual data. We also train several variants of Transformer such as DynamicConv [20] and Transformer with relative position representations. We select back-translated data by length and alignment features. We average the parameters of several best checkpoints [3] in a single training process to get a better single model. Translation models trained on mixed data are fine-tuned on real data provided by the evaluation organizer. Finally, we translate source texts by ensemble several best performing models and rerank the n-best lists with K-batched MIRA algorithm [4].

With above techniques, our system evaluated with BLEU [13] improved for a large margin. We also tried a few methods used in other neural machine translation systems without seeing significant improvements.

© Springer Nature Singapore Pte Ltd. 2020
J. Li and A. Way (Eds.): CCMT 2020, CCIS 1328, pp. 76–82, 2020.
https://doi.org/10.1007/978-981-33-6162-1_7

2   Machine Translation System

Since there is far less parallel data for Uighur → Chinese translation, we adopt several effective techniques for alleviating data starvation problem. The following sections describe how we build a well-performing system for Uighur → Chinese translation in low-resource scenario.

2.1   Pre-processing

Table 1. Statistics of pre-processed parallel data.

Translation direction	#Sentence pairs
Uighur → Chinese	165792
Uighur → Chinese (sample 1)	6340403
Uighur → Chinese (sample 2)	6340804

We escape special characters and normalize punctuation characters with Moses [10][1]. Then we tokenize sentences for Chinese with pkuseg [12][2]. Sentences with more than 100 words were removed for both Uighur and Chinese. We also filter parallel data where Chinese sentence is 6 times longer than Uighur sentence or Uighur sentence is 4 times longer than Chinese. We learn word alignment with fast_align [6][3] and filter sentence pairs whose alignment rates are less than 0.6. The statistics of pre-processed parallel data are shown in Table 1. The remaining data is processed by Byte Pair Encoding [15][4], with 32K merge operations for both Uighur and Chinese.

2.2   Architecture

Table 2. Architecture hyper-parameters of Transformer Big in our system.

Hyper-parameter name	Hyper-parameter value
Embedding size	1024
Hidden size	1024
Ffn inner size	4096
Attention heads	16
Dropout	0.2
Label smoothing	0.1

---

[1] https://github.com/moses-smt/mosesdecoder.
[2] https://github.com/lancopku/pkuseg-python.
[3] https://github.com/clab/fast_align.
[4] https://github.com/rsennrich/subword-nmt.

We adopt Transformer Big as our base model and tune a few architecture hyper-parameters in current setting, which are shown in Table 2. We train all models by optimizing cross entropy loss with label smoothing. Adam optimizer [9] ($\beta 1 = 0.9$, $\beta 2 = 0.98$, $\epsilon = 10^{-9}$) was used for optimization. Learning rate is linearly increased during the first 4000 steps, and then decreased with inverse square root function of steps as in [19]. We train all models on 4 NVIDIA Tesla V100 GPUs.

To obtain more diversed models for ensembling, we train two variants of vanilla Transformer: Transformer with relative position representations [16] (Relative Transformer) and DynamicConv [20]. Checkpoint averaging [3] is also used to get a stronger model.

2.3 Back-Translation of Monolingual Data

Back-translation has been proved as an effective method for data augmentation of neural machine translation [7,14], especially in low-resource scenarios. With only 165K provided parallel data, Transformer Big performs worse than Transformer Base, seeing Table 3. We train a Chinese → Uighur translation model, taking Transformer Base architecture. Then we apply the trained Transformer to translate large scale monolingual sentences in Chinese to Uighur and construct pseudo Uighur → Chinese translation parallel data.

Table 3. Back-translation with different strategies

Setting	BLEU
Transformer Base w\o BT	38.56
Transformer Big w\o BT	37.01
Transformer Big w\BT (beam search)	45.27
Transformer Big w\BT (top-10 sampling)	45.34
Transformer Big w\BT (top-10 sampling) + tag	46.00

We experiment with several methods to generate synthetic data as proposed in [7], such as beam search and top-k sampling. We find top-k sampling is more effective as shown in Table 3. A possible explanation is that top-k sampling introduce moderate noise into synthetic data, which makes pseudo data generated by top-k sampling contain stronger training signal [8].

It is useful to distinguish real data and synthetic data during training since synthetic data is usually more noised. A simple method distinguish real data and synthetic data is adding a tag in front of each sentences, which is called Tagged Back-Translation [2]. Experimental results in Table 3 proved its effectiveness in Uighur → Chinese translation.

We construct two synthetic datasets (named sample1 and sample2) by top-10 sampling in back-translation and filter sentence pairs with length and alignment features.

2.4 Fine-Tuning

Table 4. Fine-tuning trained models on real data

Model	Before tuning	After tuning
Transformer (sample 1)	46.00	46.96
Transformer (sample 2)	45.32	46.76
Checkpoint averaging (sample 1)	45.53	47.26
Relative transformer (sample 1)	45.56	47.53
Relative transformer (sample 2)	45.40	47.43
Dynamic convolution (sample 1)	45.42	46.82

There is domain divergence between real data and synthetic data, since synthetic sentence pairs are in general domain while real data specific in news domain. We fine-tune translation models trained on mixed data on real data to adapt them specific to target domain.

As indicated in Table 4, fine-tuning trained model on real data boost the performance of translation models for a large margin evaluated by BLEU scores on development set.

2.5 Ensemble Translation

Many literatures [1,18] have shown the effectiveness of ensemble learning for improving translation quality. We translate evaluation source texts by ensembling several diversed and best performing models. Our experimental results in Table 5 present stable increments of translation quality with ensembling more best performing models.

Table 5. Ensemble translation: index i means the i-th model in Table 4

Ensemble selection	BLEU
4 + 5 (beam size = 5)	47.97
3 + 4 + 5 (beam size = 5)	48.21
1 + 3 + 4 + 5 (beam size = 5)	48.51
1 + 3 + 4 + 5 + 6 (beam size = 5)	48.54
1 + 2 + 3 + 4 + 5 + 6 (beam size = 5)	48.63
1 + 2 + 3 + 4 + 5 + 6 (beam size = 24)	48.80

2.6   Reranking

We generate the n-best translation lists by ensembling 6 best performing models with beam size = 24. We hand-craft several features for reranking the n-best lists, including log probability of each single translation model, target-to-source translation score, right-to-left translation score [11], n-gram language model perplexity[5] and beam index. The reranking model is tuned by K-batched MIRA algorithm [4]. BLEU score evaluated on development set achieves 49.17 after reranking.

3   Results

Table 6 shows our systems evaluated by BLEU on development set. For Uighur → Chinese translation, BLEU scores [13] are computed at character level. For the last 4 rows, each model is based on the model described in the previous row.

Table 6. Translation quality evaluated by BLEU on development set

System	Uighur → Chinese
Transformer Base	38.56
Transformer Big	37.01
+ Back Translation	46.00
+ Fine-tuning	46.96
+ Ensembling	48.80
+ Reranking	49.17

We can see that back-translation, fine-tuning, ensemble translation and reranking consistently boost the performance of the Uighur → Chinese translation system. During these techniques, back-translation is most effective in low-resource scenario.

4   Conclusion

This paper presents our submission for CCMT-2020 Uighur → Chinese translation task. We obtain a strong baseline system by tuning Google's Transformer Big architecture and continually improve it by back-translation, fine-tuning, ensembing and reranking.

---

[5] https://github.com/kpu/kenlm.

References

1. Bahdanau, D., Cho, K., Bengio, Y.: Neural machine translation by jointly learning to align and translate. In: ICLR 2015: International Conference on Learning Representations 2015 (2015)
2. Caswell, I., Chelba, C., Grangier, D.: Tagged back-translation. In: Proceedings of the Fourth Conference on Machine Translation (Volume 1: Research Papers), pp. 53–63 (2019)
3. Chen, H., Lundberg, S., Lee, S.I.: Checkpoint ensembles: ensemble methods from a single training process. arXiv preprint arXiv:1710.03282 (2017)
4. Cherry, C., Foster, G.: Batch tuning strategies for statistical machine translation. In: Proceedings of the 2012 Conference of the North American Chapter of the Association for Computational Linguistics: Human Language Technologies, pp. 427–436 (2012)
5. Devlin, J., Chang, M.W., Lee, K., Toutanova, K.: Bert: pre-training of deep bidirectional transformers for language understanding. In: NAACL-HLT 2019: Annual Conference of the North American Chapter of the Association for Computational Linguistics, pp. 4171–4186 (2019)
6. Dyer, C., Chahuneau, V., Smith, N.A.: A simple, fast, and effective reparameterization of IBM model 2. In: Proceedings of the 2013 Conference of the North American Chapter of the Association for Computational Linguistics: Human Language Technologies, pp. 644–648 (2013)
7. Edunov, S., Ott, M., Auli, M., Grangier, D.: Understanding back-translation at scale. In: EMNLP 2018: 2018 Conference on Empirical Methods in Natural Language Processing, pp. 489–500 (2018)
8. Hu, B., Han, A., Zhang, Z., Huang, S., Ju, Q.: Tencent minority-mandarin translation system. In: Huang, S., Knight, K. (eds.) CCMT 2019. CCIS, vol. 1104, pp. 93–104. Springer, Singapore (2019). https://doi.org/10.1007/978-981-15-1721-1_10
9. Kingma, D.P., Ba, J.L: Adam: a method for stochastic optimization. In: ICLR 2015: International Conference on Learning Representations 2015 (2015)
10. Koehn, P.: Open source toolkit for statistical machine translation. In: Proceedings of the 45th Annual Meeting of the Association for Computational Linguistics Companion Volume Proceedings of the Demo and Poster Sessions, pp. 177–180 (2007)
11. Liu, L., Utiyama, M., Finch, A.M., Sumita, E.: Agreement on target-bidirectional neural machine translation. In: 2016 Conference of the North American Chapter of the Association for Computational Linguistics: Human Language Technologies, NAACL HLT 2016 - Proceedings of the Conference, pp. 411–416 (2016)
12. Luo, R., Xu, J., Zhang, Y., Ren, X., Sun, X.: PKUSEG: a toolkit for multi-domain Chinese word segmentation. arXiv preprint arXiv:1906.11455 (2019)
13. Papineni, K., Roukos, S., Ward, T., Zhu, W.J.: BLEU: a method for automatic evaluation of machine translation. In: Proceedings of 40th Annual Meeting of the Association for Computational Linguistics, pp. 311–318 (2002)
14. Sennrich, R., Haddow, B., Birch, A.: Improving neural machine translation models with monolingual data. In: Proceedings of the 54th Annual Meeting of the Association for Computational Linguistics (Volume 1: Long Papers), vol. 1, pp. 86–96 (2016)
15. Sennrich, R., Haddow, B., Birch, A.: Neural machine translation of rare words with subword units. In: Proceedings of the 54th Annual Meeting of the Association for Computational Linguistics (Volume 1: Long Papers), vol. 1, pp. 1715–1725 (2016)

16. Shaw, P., Uszkoreit, J., Vaswani, A.: Self-attention with relative position represen-
    tations. In: NAACL HLT 2018: 16th Annual Conference of the North American
    Chapter of the Association for Computational Linguistics: Human Language Tech-
    nologies, volume 2, pp. 464–468 (2018)
17. Srivastava, N., Hinton, G., Krizhevsky, A., Sutskever, I., Salakhutdinov, R.:
    Dropout: a simple way to prevent neural networks from overfitting. J. Mach. Learn.
    Res. **15**(1), 1929–1958 (2014)
18. Sutskever, I., Vinyals, O., Le, Q.V.: Sequence to sequence learning with neural
    networks. In: Advances in Neural Information Processing Systems, vol. 27, pp.
    3104–3112 (2014)
19. Vaswani, A.: Attention is all you need. In: Proceedings of the 31st International
    Conference on Neural Information Processing Systems, pp. 5998–6008 (2017)
20. Wu, F., Fan, A., Baevski, A., Dauphin, Y.N. and Auli, M.: Pay less attention with
    lightweight and dynamic convolutions. In: ICLR 2019: 7th International Conference
    on Learning Representations (2019)

Description and Findings of OPPO's Machine Translation Systems for CCMT 2020

Tingxun Shi, Qian Zhang, Xiaoxue Wang, Xiaopu Li, Zhengshan Xue[✉],
and Jie Hao

Manifold Lab, OPPO Research Institute, Beijing, China
{shitingxun,zhangqian666,wangxiaoxue,lixiaopu,
xuezhengshan,haojie}@oppo.com

Abstract. This paper demonstrates our machine translation systems for the CCMT 2020, which is composed of four parts. The last three parts report our results in the contest, each respectively focuses on English-Chinese bi-direction translation, Japanese-Chinese-English multi-lingual translation (patent domain), and Chinese minority languages to Mandarin Chinese translation. In each part, we will demonstrate our work on data pre-processing, model training as well as the application of general techniques, such as back-translation, ensemble and reranking. Besides, during our experiments, we surprisingly found that simply applying different Chinese word segmentation tools on low-resource corpora could bring obvious benefit across different tasks, and we will separate an independent section to discuss this finding. Among the 7 directions we participated in, we ranked the first in 6 tasks (For the corpus filtering task, we ranked first in the 500 million words sub-task) and the second for the rest.

Keywords: Back-translation · Ensemble · Reranking · Multi-task learning

1 Introduction

Machine translation has always been a popular research field in the Natural Language Processing (NLP) area. In recent years, Transformer [1]-based system has become the main-stream architecture for the neural machine translation tasks, brought the field to a new stage. Since the results generated by the model are more promised, and the system is generally end-to-end, no longer as complex as the systems in the statistical machine translation era, some new research problems have emerged, such as low-resource translation, multi-lingual translation, and so on. This report describes our (OPPO's) machine translation system designed for the 16th. China Conference on Machine Translation (CCMT 2020), including all the models we trained for nearly all tasks. These tasks could be further divided into three categories, which are:

J. Li and A. Way (Eds.): CCMT 2020, CCIS 1328, pp. 83–97, 2020.
https://doi.org/10.1007/978-981-33-6162-1_8

- English ↔ Chinese bi-directional translation. This task provides a great amount of parallel corpus which can be used to train a good enough Transformer model. We combined rule-based and model-based preprocessing methods to clean the corpus and further experimented with some other well-known techniques, such as back-translation, domain adaptation, knowledge distillation, and reranking, and the results show that they can all generally improve the translation quality more or less. The models trained in this task are also utilized in the parallel corpus filtering task to score the sentence pairs.
- Japanese → English translation in the patent domain. This is the second year we participate in this task and different from the one we proposed in the last year [2], in this paper we will show a new solution, which is based on multi-lingual translation.
- China's Minority Languages (including Uighur, Tibetan and Mongolian (in traditional form). Written as "minority languages" below for short) to Mandarin Chinese (written as "Mandarin" below for short). All of these three tasks are resource limited, but experiments show that both the model architecture and the extra boosting techniques applied in the English ↔ Chinese section are also applicable in the low-resourced tasks. Furthermore, we surprisingly found a simple extra preprocessing to the corpus can bring a big gain for the model. We will give a brief introduction to this preprocessing method in the corresponding section, and consider to publish an individual paper to explore its application scope.

As this report introduces multiple different systems together, and most of them share a similar data processing way, model architecture and improving techniques, to avoid duplicated words, we will demonstrate the common, general skills firstly, i.e. in the English ↔ Chinese translation system in Section Three (English → Chinese corpus filtering task is also described here). Section Four shows the Japanese → English patent translation system and in Section Five this paper introduces our system for the China's minority languages to Mandarin translation task. The Final Section will summary this report and list our further work. What's more, we will make a space for our finding during the contest, show how did we combine different word segmentation results for Chinese in low-resource translation tasks to improve the system.

2 Applying Multiple Word Segmentation Tools

As written Mandarin does not have explicit word boundaries, research on the segmentation methods of Mandarin has been always an active field in Chinese NLP [21,22]. This also leads to the development of some well-known Chinese segmentation tools such as *jieba*, *pkuseg*, and so on. Generally, people use a single segmentation tool in their experiments, and besides this mainstream way, other works like [12] argue that translating from pure character-based data can also reach a SOTA result.

Different from the current practices, inspired by the concept of multiple tasks transfer learning, in this paper we propose a new way to handle how the Chinese words should be segmented. For a given sentence pair, we segmented the Chinese side by two different tools, *jieba*[1] and *pkuseg* [13], then combined the segmented results together with another result that is from simply splitting the Chinese sentence into characters. Since after such a process one sentence pair becomes three, we added a tag in front of both the source side and the target side, indicating for the current pair which segmentation tool is applied. In this way, the size of parallel corpus is augmented to three times bigger (for Uighur → Mandarin task, we additionally segmented the Mandarin corpus using *scws*[2], so the corpus is four times bigger).

Furthermore, as Mandarin doesn't have explicit word boundaries, we decided to remove the BPE suffices "@@" for all subwords. The reason is in some cases, the subword generated by BPE tools actually share the same meaning from the corresponding independent word, the only literal difference is the former one has an extra BPE suffix. For example, suppose a segmentation tool sees "国际贸易" (international trade, "国际" means "international" and "贸易" means "trade") as a whole word, and this word is divided into "国际 @@ 贸易" by BPE tools. In this case, the two different tokens "国际 @@" and "国际" actually have the exactly same meaning, using different tokens to distinguish them is unnecessary, and the extra introduced token enlarges the vocabulary, makes the network bigger, thus is easier to be overfit in the low-resource tasks.

After having removed the BPE suffices, we iterated all subwords again: for a given subword which has been removed the suffix, if there isn't a same full word existing in the vocabulary, then we shatter it again into characters. The intuition behind such a decision is we think in this way model can learn more information from the character-level corpus.

Table 1 shows the effect after having applied multiple word segmentation tools on Tibetan → Mandarin task. The different techniques we listed above are all introduced step by step, thus this table can be seen as the result of ablation analysis. From the table we can see all the introduced techniques helps to boost the system, so the root cause that brings the improvement would not be simple augmenting the dataset. A further analysis would be carried on in our future work.

As a sentence can be translated into different results according to the label in the very beginning, we can extract different results by the segmentation tags, compare there BLEU scores on both the validation set and the online test platform. For Tibetan task, we found using a statistical language model (*kenlm* model) to evaluate the candidates, picking out the one has the highest score, can achieve more gains (For Uighur and Mongolian this method fails. Nevertheless we submit our final results according to the reranking system, so this does not matter much).

We also tried to apply this method on the English → Chinese task, which has abundant training corpus, but did not affect the system (neither improved

---

[1] https://github.com/fxsjy/jieba.
[2] http://www.xunsearch.com/scws/.

Table 1. Improvement achieved by using multiple segmentation tools on Tibetan →
Mandarin tasks. Here the online test data is actually the CCMT 2019 test dataset.
Scores reported on the validation set are calculated by SacreBLEU (character-level
BLEU4), on the online test are calculated by the official evaluation suite (character-
level BLEU5-SBP)

Method	Validation set BLEU	Online test BLEU
Baseline model (Character-based Mandarin)	44.2	54.74
+ *pkuseg* segmented Mandarin	45.4	(not tested)
+ Multiple segmentation, without segmentation tag	45.7	(not tested)
+ segmentation tag & keeping BPE symbol	46.1	(not tested)
+ removing BPE symbol	46.2	55.90
+ selected by *kenlm*	**46.7**	**56.69**

nor harmed the system). This results shows that our proposed method is more
applicable in the low-resource scenarios.

3 English ↔ Chinese Machine Translation Task

In the neural machine translation era, models become much bigger, contains
more parameters, therefore generally requires more data for training. The CCMT
2020 English ↔ Chinese task provides the biggest parallel dataset across all
translation tasks held in CCMT 2020, contains roughly 28 million parallel sentence pairs and another 20 million official released forward-translated data[3], such
a big data amount gave us confidence to train an applicable model and experiment with some other techniques. What's more, as models generally need high
quality data to generate more promising results, we also designed a data preprocessing and filtering pipeline to clean the data. This pipeline was also reused in
the other tasks that will be presented later.

3.1 Data Preprocessing

Our data preprocessing procedure can be divided into two parts: In the **preprocessing** part, sentences are normalized, generally including symbol normalization, tokenization, word segmentation (for the languages that don't have explicit
words boundaries, such as Chinese), and true casing (for the languages of which
the letters have different cases). In this part, sentences are only converted but
not dropped. The concrete steps are:

- Simplifying Chinese characters. Traditional Chinese characters are converted
  into their corresponding simplified forms.

---

[3] Including datasets released by WMT 2020, which are allowed in CCMT 2020.

- Punctuation normalization. E.g. all different hyphens are converted into the standard one. For Chinese, this step includes an extra function to convert all full-width symbols (not only punctuations, but also numbers and Latin letters) into half-width, except commonly used punctuations such as full stops, commas, question marks and exclamation marks.
- Word segmentation for Chinese. We used *pkuseg* [13] as the segmentation tool.
- Tokenization. We used *Moses*[4] as the tokenization tool.
- True casing for English. The true casing tool is also from the Moses suites

In the **filtering** part, sentence pairs that have low qualities are removed. We apply two kinds of methods to filter the corpus: For the **heuristic** methods, we set up some rules and thresholds, including

- Remove the sentence pairs that contain too many non-sense symbols.
- Remove the sentence pairs that contain too long sentences (have more than 160 words).
- Remove the sentence pairs that the count difference between numbers in Chinese side and numbers in English side is greater than or equal to 3.
- Remove the sentence pairs that the count difference between punctuations in Chinese side and punctuations in English side is greater than or equal to 5.
- Sentence pairs that have abnormal length ratio, here "length" is the count of words of a sentence. We set the upper bound of words count ratio between English and Chinese to 2.2 and the corresponding lower bound is 0.7.
- Deduplication.

The rest corpus is then filtered by **alignment** information. We used *fast_align*[5] [17] to calculate the alignment information between Chinese corpus and English. Both sentence-level and word-level alignment scores are referred to. For sentence-level information, we calculated scores from English to Chinese (noted as **enzh** below for short) and scores in the reversed direction (noted as **zhen** below for short), then averaged these two scores. For word-level information, we first averaged the two scores by the word counts of each other. Our threshold for the sentence-level alignment information is -16, and for the word-level is -2.5.

After the steps we listed above, about 17 million parallel sentence pairs and 14 million official forward-translated pairs were left. We also cleaned some official provided English & Chinese monolingual corpus for back-translation and forward-translation later, data sources and corresponding dataset size after cleaning are:

- Chinese: 10.55 million, including 7.5 million LDC data and 3 million Newscrawl data.

---

[4] https://github.com/moses-smt/mosesdecoder/blob/master/scripts/tokenizer/tokenizer.perl.

[5] https://github.com/clab/fast_align.

– English: 57 million, including 20 million Newscrawl data, 20 million LDC data and 17 million News-discuss data.

We strictly followed the requirements, built constrained systems for all the shared tasks in the CCMT 2020.

3.2 Model Training

Cleaned corpus is then used to divide words into subwords. We merged the English side and the Chinese side of the parallel corpus to train BPE separations, the count of BPE merge operations is 32K, then we built vocabulary lists for two languages independently. We applied 8 heads Transformer-Big [1] architecture to train our models, using *fairseq* [5]. For zhen task, we tried different hyperparameters to train several models for getting ensemble model: learning rates ranged from 0.0003 to 0.0008, warmup steps fixed at 16,000, dropout ranged from 0.2 to 0.3. For enzh task, the hyperparameters are all fixed (but tried different random seeds): learning rate was 0.0003, warmup steps was 15,000, feedforward network dimension was 15,000. In all the CCMT 2020 tasks we used Adam optimizer [4] to optimize the models.

During training we tried the following techniques:

– Back-translation and forward-translation. For back-translation [6] we first trained models on parallel corpus, then used these models to translate monolingual corpus of the target language, and combined the synthetic pseudo parallel corpus with the original one. Although [8] indicates that adding some noises by using sampling-based decoding can improve the results, in this task we found argmax-based beam search still performs the best. Later we found adding *forward-translation*, i.e., using models to translate monolingual corpus of the source language can also boost system's performance. We borrowed ideas from [9], used ensemble model to again back & forward translate both monolingual corpus and parallel corpus, and as [7] we did such processes for three rounds.
– Domain adaptation. We found some sentence pairs in the parallel dataset are somehow away from the test dataset: test dataset is in the news domain, however the parallel dataset provides some examples from UN conferences. Besides, style of synthetic data is also different from that of the parallel one (i.e. the "translationese issue"). To address such a problem we introduced a two-phases fine-tuning method: After having trained the model on a big, natural and synthetic dataset mixture, we first fine-tuned the model on the official provided parallel corpus only, and then fine-tuned the model again only using a tiny dataset, newstest2017, since the newstest dataset always has higher quality and fits the domain well.
– Ensemble. As presented we always train several models for the same task, using different hyperparameters and/or random seeds, to get an ensemble model. Experiments show that ensemble model in most cases can improve the result.

- Reranking. We generated several best candidates (generally 10) from the ensemble model, and scored them by several small models. The scorers include forward-translation models (e.g. the models used to compose the ensemble model), backward-translation models (e.g. the models used to generate back-translation results) and language models. For each kind of model, we also trained its right-to-left counterpart (i.e. reverse both the source sentences and target sentences then train a model) to enrich the choice of score models. We applied K-batched MIRA [11] to rerank the candidates.

The two tables below show our results achieved in both **zhen** and **enzh** tasks, with the techniques listed above (Tables 2 and 3).

Table 2. Our systems for **zhen** translation task. Scores are reported on the **newstest2019** dataset and evaluated by SacreBLEU [20]. Scorers for reranking are composed of 3 forward left-to-right (l2r) models, 3 forward right-to-left (r2l) models, 3 backward r2l models and 2 l2r Transformer language models.

System	BLEU	Absolute improvement	Relative improvement
Baseline (trained by parallel corpus only)	28.8	–	–
+ back-translation	29.8	+1.0	+1.0
+ forward-translation	34.5	+5.7	+4.7
+ fine-tuned by newstest2017	36.7	+7.9	+2.2
+ ensemble & reranking	38.3	+9.5	+1.6

Table 3. Our systems for **enzh** translation task. Scores are reported on the **newstest2019** dataset and evaluated by SacreBLEU. We also tried the two-phases fine-tuning on the models trained by adding forward-translation, but no gains observed (So the "relative improvement" given in the "forward-translation" row is calculated based on the "back-translation" row). Scorers for reranking are composed of 5 forward l2r models, 3 forward r2l models, 3 backward r2l models and 3 l2r Transformer language models.

System	BLEU	Absolute improvement	Relative improvement
Baseline (trained by parallel corpus only)	38.6	–	–
+ back-translation	39.1	+0.5	+0.5
+ fine-tuned by parallel corpus	40.6	+2.0	+1.5
+ fine-tuned by newstest2017	41.3	+2.7	+0.7
+ forward-translation	41.9	+3.3	+2.8
+ ensemble	42.7	+4.1	+0.8
+ reranking	43.2	+4.6	+0.5

3.3   Corpus Filtering Task

This year we also participated in the Chinese $\leftrightarrow$ English corpus filtering task. This task requires us to score every sentence pair in a given parallel corpus. We used the models trained for the Chinese $\leftrightarrow$ English translation task to score the sentence pairs, averaged the score got by the forward model and backward model, then sorted them according to the averaging score. In the contrast system, we selected out the sentence pairs that contain traditional Chinese characters, and put all of them in the end of the submission.

4   Japanese $\rightarrow$ English Translation Task (Patent Domain)

The Japanese (ja) $\rightarrow$ English (en) translation task in the patent domain is designed as a multi-lingual, zero-shot, domain-specific task. Official data doesn't contain any jaen parallel dataset, but is composed of two independent datasets: one is Japanese $\leftrightarrow$ Chinese (jazh), the other is English $\leftrightarrow$ Chinese (enzh). Based on the system we proposed in the last year [2], we made some further improvements, including introducing a step inspired by the mainstream multi-lingual translation solutions.

4.1   Data Preprocessing

We adopted exactly the same data preprocessing and filtering as we described in the enzh section, including the hyperparameter settings, with an extra step to convert all CJK characters in the Japanese corpus to Japanese *kanji* forms. Both of the officially provided datasets contain 3 million sentence pairs, after cleaning jazh corpus had 2.9 million pairs left, and enzh had 2.8 million. We used *pkuseg* to segment Chinese sentences, *mecab*[6] to segment Japanese sentences, and *Moses-tokenizer* to tokenize sentences. All source and target side corpora are mixed to train the BPE subword segmentation, merging operations were applied 32 K steps but we did not share the vocabulary among the source and the target.

4.2   Model Training

A direct solution from the given data is to train two systems, one is from Japanese to Chinese, the other is from Chinese to English. The key defect of this system is, all the source sentences will be translated by two models consecutively. As currently models are not able to guarantee the quality of the generated results, each step has a probability to make mistakes. What's the worse, consecutive translating may even have the risk to augment the wrong signals.

In the solution we proposed last year [2], we built a zhja system, and translated the Chinese sentences in zhen corpus into Japanese, therefore we can get a pseudo Japanese $\rightarrow$ English parallel dataset which contains 2.8 million pairs

---

[6] https://taku910.github.io/mecab/.

of data. This year, as we saw a success of forward-translation in English ↔ Chinese task, we applied the same processing way here, built a `zhen` system, thus made another 2.9 million synthetic pairs. Combined the two synthetic datasets together we can get a corpus of which the size is 5.7 million. We first trained a Transformer-Big model on this "raw" dataset, then followed the model-based filtering process described in [3], used the same model to score all the sentence pairs and further removed 100k sentences away. We trained several different checkpoints, mainly different in the learning rate (range from 0.0003 to 0.0008). The warmup steps was fixed at 16,000. Among the checkpoints we got, the best model's BLEU on the validation set is 39.5 (evaluated by SacreBLEU[7], reported on the character-level), and the ensemble model's score is 41.0.

After having built the synthetic `jaen` corpus, we tried a multi-lingual machine translation method inspired by [16]: We combined this synthetic dataset along with the other two officially released corpora, added labels at the beginning of the sentences to indicate the concrete translation directions, then trained several models on this mixed multi-lingual dataset. This multi-lingual system improved the single model for one point, from 39.5 to 40.5. However, the ensemble model only got a 0.1 point gain. Furthermore, reranking by K-Batched MIRA also brought a 0.4 point improvement.

The overview of our system for patent domain multi-lingual `jaen` translation task is listed in Table 4. We also tried some fine-tune methods but didn't see any positive results. Sometimes the score on the validation set was extremely high but by analyzing the generated results we found the model actually overfitted severely, one concrete phenomenon we observed is the fine-tuned model (after decades of epoch) always add an extra "the" before countries' names (e.g. "the China"), which obviously breaks grammar rules of English. As the corpus we used to fine-tune models are selected from validation dataset according to test dataset by fda algorithm [19], we doubt there exists some gap between the validation set and test set.

Table 4. Our systems for patent domain multi-lingual `jaen` translation task. Scores are reported on the validation set and evaluated by SacreBLEU. Baseline score is from the system we designed in the last year, differs from the current system in two aspects: 1. The evaluator applied in the last year is multi-bleu, 2. The Chinese segmentation used last year is *jieba*

System	BLEU	Absolute improvement	Relative improvement
Baseline (no forward-translation)	37.8	–	–
+ forward-translation	39.5	+1.7	+1.7
+ multi-lingual processing	40.5	+2.7	+1.0
+ ensemble	41.1	+3.3	+0.6
+ reranking	41.5	+3.7	+0.4

---

[7] https://github.com/mjpost/sacrebleu.

5   Minority Languages → Mandarin Translation Task

Comparing with the English ↔ Chinese translation task, datasets released for the minority languages → Mandarin translation task are relatively much smaller (Details can be found in Table 5). Training a good deep neural network model on such low-resource datasets brings a bigger challenge to us, therefore some extra processing steps are introduced.

Table 5. Corpora sizes in the China's minority languages → Mandarin translation task. The statistics information is collected from the very original, raw **training** data, so all of the sentences in the source side are not tokenized. As Tibetan does not show the word boundary explicitly neither (as Mandarin), in the corresponding row we count the characters amount for Tibetan

Language pairs	# Sentence pairs	# Tokens in the source side	# Characters in the target side
Uighur → Mandarin	169,525	3,114,647	17,244,943
Tibetan → Mandarin	162,096	32,158,312	7,839,757
Mongolian → Mandarin	261,454	5,993,512	24,579,256

5.1   Data Preprocessing

Small data amount is a double-edged sword: From one side it makes training a good model more difficult, but from the other side it allows us to do a more careful cleaning. As the dataset is small, model is more vulnerable, easier to be disturbed by noises, so a careful cleaning is necessary and even more important. For each language pair, we list all the characters in the raw training dataset, and according to the character list we further design ad-hoc rules to modify low-frequency, irregular characters. The rules can be roughly divided into below categories:

1. Symbol forms unification: For a given symbol/punctuation, map its all variations to the most popular one (in most cases, to the corresponding ASCII form). For example, "EM Dash" (Unicode 0x2014) is mapped to "dash" (Unicode 0x002e).
2. Conversion between full width symbols and half width symbols. In the source side (i.e. for all minority languages) full width symbols are changed to half width symbols, and in the target side (i.e. for all Mandarin corpus) some common half width symbols are converted to their full width counterparts (such as full stops, commas).
3. Removal of the invalid/invisible/unnecessary characters. For example, Unicode 0xe5e7 is removed.

Special Process for Tibetan. In the given three different source languages, Tibetan, similar to Mandarin Chinese, doesn't have explicit word boundaries, which is different from Uighur and Mongolian. Since we didn't find an ideal Tibetan word segmentation tool released by the domestic team, we chose to train a "character"-based model[8] for Tibetan → Mandarin task. The extra process for Tibetan contains

- Remove all *initial yig mgo mdun ma*s (Unicode `0x0f04`).
- Replace all morpheme delimiters (*tseg*, Unicode `0x0f0b`) to spaces.
- Add spaces around all full stops (*tshig-grub*s, Unicode (`0x0f0d`)) and roof over brackets (both *ang khang g.yon*s and *ang khang g.yase*s, Unicode `0x0f3c` and `0x0f3d`).

After this character-based cleaning, the following preprocessing steps are similar to the ones described in Sect. 2, which contains two stages: In the first stage some normalization methods are applied, such as space normalization, punctuation normalization, symbol unescaping, and (for Mandarin) simplifying traditional Chinese characters. The second stage involves some rule-based filtering steps, like deduplication, language identification, statistical information based filtering (including count of words/characters, source-target sentences length ratio, ratio of letters for each sentence, ratio between count of characters and count of words for each sentence), and alignment based filtering. We again used *fast_align* to get the alignment information of each training dataset[9]. Notice here that in the statistical information based filtering and alignment based filtering, for each item we didn't manually hard-code the concrete thresholds, but indicated percentiles respectively. The most frequently used percentiles are 0.1% and 99.9%, since we want to keep as many data pairs as possible, meanwhile also need to get rid of the real abnormal ones (for example, too long sentences). For Mandarin, we take an extra filtering step: use the 3,500 common used Chinese characters list and a kenlm language model [10] trained on officially provided monolingual Mandarin corpus to filter out sentences that contain too many irregular codes.

Table 6 shows the data processing results after the filtering. As [15] indicated, we also found the official validation set for Tibetan task has a low quality, so we fully discarded it and sampled 1,440 sentences from the training set as new validation set. BPE subwords are applied to all the three tasks and are all trained separately. For Uighur and Tibetan the BPE merge operations 32K and for Mongolian it is 16K. As we showed in Sect. 2, we applied multiple word segmentation tools on all the three tasks.

---

[8] More accurately, morpheme-based model.

[9] For Uighur → Mandarin task we didn't filter the corpus according to alignment information, since we find sometimes a Mandarin word can be a long phrase in Uighur. e.g. "法治" (rule of law) is officially translated to *"qanun arqiliq idare qilish"*. (Uighur here is transliterated by Uighur Latin alphabet (ULY)). For Tibetan, alignment information is calculated on a character-level corpus, means not only the Tibetan data is segmented by morpheme, but also the Mandarin data is split into characters.

Table 6. Corpus filtering information for Minority → Mandarin tasks

Language pairs	# Raw sentence pairs	# Kept sentence pairs	Retention rate
Uighur → Mandarin	169,525	163,762	96.60%
Tibetan → Mandarin	162,096	147,440	90.96%
Mongolian → Mandarin	261,454	228,225	96.18%

5.2 Model Training

Before discovering the multiple word segmentations method (described in Sect. 2), the baseline models in this section (for Uighur and Tibetan tasks) were trained by *marian*. We adopted the Transformer-Big architecture described in [1]. The optimizer we used is Adam, learning rate was set to 0.0001, warmup was set to 16,000, gradient norm clipping was set to 5. We also applied label smoothing, the corresponding parameter is 0.1. When evaluating the model on the validation set, the beam search size was 24 and the normalization is 1.5. We set the early stopping validation counts to 50. We followed [15], didn't average the checkpoints, but used a smoothing averaging method with the factor set to 10^{-4}.

Since multiple segmentation method is validated to be effective, we applied it on all the three Minority language tasks, using *fairseq* to train the models. The reason we switched the framework is that we found when dataset becomes larger, models trained by *fairseq* with our frequently used configuration is slightly better than those trained by marian. The changes for our *fairseq* configuration are:

- gradient norm clipping set to 0.1
- gradient update frequency set to 8
- dropout set to 0.3 (ReLU dropout and attention dropout are kept as 0.1)
- warmup initial learning rate set to 10^{-7}
- beam search for decoding in validation is set to 5, and length penalty is set to 2

What's more, we didn't adopt the smoothing averaging checkpoints as we did when using marian. Here we did not specify what learning rate and warmup steps we applied, because they actually vary across different tasks, and even in the same task we tried different configurations to get different checkpoints to further compose the ensemble model. Generally, the most common used combination is learning rate 0.001 and warmup step 16,000, but per our experiences learning rate can range from 0.0008 to 0.002, and warmup can range from 8,000 to 32,000. Sometimes bad combination can lead to gradient explosion, but mostly if the training converges, the result could be acceptable (after averaging checkpoints, the score difference between the best one and the worst one is less than 1 BLEU).

As discussed in the previous section, we found using synthetic data generated by back-translation can improve the system a lot. In the Minority → Mandarin tasks, the same process is also applied in all the three directions. The Mandarin corpus is again segmented in three different ways (character, *jieba* and *pkuseg*. For Uighur again plus *scws*). For the back-translation model, we cannot guarantee the same sentence segmented in different ways can be translated into the same results, but we did not process the back-translated results, leaving the divergence and hoped this could be helpful noise for the model. However, it could be better to take some extra experiments to see how unified version of back-translated results can affect the system.

Table 7 shows our three Minority → Mandarin systems results, showing the gains brought by each technique. Besides the multiple segmentation methods, we applied roughly the similar techniques we described in the previous English ↔ Chinese section, including back-translation, domain adaptation, ensemble and reranking. It should be noted that for the Minority → Mandarin tasks, monolingual data is only available for Mandarin, so we were not able to do forward-translation using our trained systems, but we followed [9] to translate back-translated source corpus using ensemble model again (ensemble knowledge distillation, ensemble KD), to augment the dataset. For back-translated data, we added a special tag `<bt>` in front of both source side and target side. For the translated results from the original data, we added a special tag `<kd>` and for the results from back-translated data, the corresponding tag is `<btkd>`. We didn't follow [15] to fine-tune our systems on knowledge distilled data, but mixed the knowledge distilled data with the original parallel corpus and back-translated data together, and trained models from the scratch.

Table 7. Overall for the minority languages → Mandarin systems. Every score is character-level BLEU calculated on the validation dataset by SacreBLEU (for Tibetan we used a part separated from the training data which contains 1440 examples, not the official validation set). Baseline for the Uighur task was trained without applying multiple segmentation tools. Reranking for the Mongolian system followed noisy channel reranking [18], the other two used K-batched MIRA. For the Uighur system, scorers contain 22 forward l2r models, 3 backward l2r models, 3 forward r2l models, 7 forward l2r Transformer language models, 7 backward r2l Transformer language models, 1 l2r *kenlm* language model and 1 r2l *kenlm* language model. For the Tibetan system, the count of forward l2r models in the scorers pack is 16, other options have the same amount as we used for the Uighur task.

System	Uighur	Tibetan	Mongolian
Baseline	38.6	46.7	61.4
+ Back-translation & Ensemble KD	48.6 (+10, +10)	47.9 (+1.2, +1.2)	63.9 (+2.5, +2.5)
+ Fine-tune on original parallel corpus	49.0 (+10.4, +0.4)	50.0 (+3.3, +2.1)	66.9 (+5.5, +3.0)
+ Model ensemble	49.4 (+10.8, +0.4)	53.0 (+6.3, +3.0)	69.5 (+8.1, +2.6)
+ Reranking	49.5 (+10.9, +0.1)	53.0 (+6.3, +0.0)	73.0 (+11.6, +3.5)

6 Conclusion and Future Work

In this report we summarized all the systems we designed for CCMT 2020. We found applying forward-translation together with traditional back-translation can bring other gains, and verified the effect of multi-lingual model training methods in the zero-shot multi-lingual task. Fine-tuning (domain adaptation) is proved to be effective if the domain of test data mismatches the domain of training data (This is again verified in the minority languages tasks, in which the Tibetan corpus is in the government domain, the Mongolian corpus is in the daily domain. However, official provided Chinese corpus is in the news domain, so fine-tune on their each parallel corpus can improve a lot). Our systems generally achieved good results: among the 7 directions we participated in, we ranked 2nd in the Mongolian → Mandarin direction with a gap of 1.3 BLEU, and 1st in the rest.

During preparing the final systems for the competition we found applying multiple Chinese segmentation tools on the low-resource dataset can boost the models' performance, we'll research on this topic further to verify whether the same idea can be also useful for other languages which have the similar feature (i.e. no explicit word boundaries), e.g. Japanese, Vietnamese, Thai and so on, and try to find a way to extend such method to languages that have explicit word boundaries. Besides, we are also interested in how to design a more effective multi-lingual translation system in the zero-shot/few-shot scenario.

References

1. Vaswani, A., et al.: Attention is all you need. In: Advances in Neural Information Processing Systems (NeurIPS 2017), pp. 5998–6008 (2017)
2. Xue, Z., Zhang, Q., Li, X., Dang, D., Zhang, G., Hao, J.: OPPO machine translation system. In: Proceedings of the 15th China Conference on Machine Translation (CCMT 2019) (2019). (in Chinese)
3. Zhang, Q., et al.: OPPO's machine translation system for the IWSLT 2020 open domain translation task. In: Proceedings of the 17th International Conference on Spoken Language Translation (IWSLT 2020), pp. 114–121, July 2020
4. Kingma, D.P., Ba, J.: Adam: a method for stochastic optimization. arXiv preprint arXiv:1412.6980 (2014)
5. Ott, M., et al.: fairseq: a fast, extensible toolkit for sequence modeling. In: Proceedings of the 2019 Conference of the North American Chapter of the Association for Computational Linguistics (Demonstrations), pp. 48–53, June 2019
6. Sennrich, R., Haddow, B., Birch, A.: Improving neural machine translation models with monolingual data. In: Proceedings of the 54th Annual Meeting of the Association for Computational Linguistics (Volume 1: Long Papers) (ACL 2016), pp. 86–96, August 2016
7. Hoang, V.C.D., Koehn, P., Haffari, G., Cohn, T.: Iterative back-translation for neural machine translation. In: Proceedings of the 2nd Workshop on Neural Machine Translation and Generation (NMT 2018), pp. 18–24, July 2018
8. Edunov, S., Ott, M., Auli, M., Grangier, D.: Understanding back-translation at scale. In: Proceedings of the 2018 Conference on Empirical Methods in Natural Language Processing (EMNLP 2018), pp. 489–500 (2018)
9. Freitag, M., Al-Onaizan, Y., Sankaran, B.: Ensemble distillation for neural machine translation. arXiv preprint arXiv:1702.01802 (2017)

10. Heafield, K., Pouzyrevsky, I., Clark, J.H., Koehn, P.: Scalable modified Kneser-Ney language model estimation. In: Proceedings of the 51st Annual Meeting of the Association for Computational Linguistics (ACL 2013) (Volume 2: Short Papers), pp. 690–696, August 2013

11. Cherry, C., Foster, G.: Batch tuning strategies for statistical machine translation. In: Proceedings of the 2012 Conference of the North American Chapter of the Association for Computational Linguistics: Human Language Technologies (NAACL-HLT 2012), pp. 427–436, June 2012

12. Li, X., Meng, Y., Sun, X., Han, Q., Yuan, A., Li, J.: Is word segmentation necessary for deep learning of Chinese representations? In: Proceedings of the 57th Annual Meeting of the Association for Computational Linguistics (ACL 2019), pp. 3242–3252, July 2019

13. Luo, R., Xu, J., Zhang, Y., Ren, X., Sun, X.: PKUSEG: a toolkit for multi-domain Chinese word segmentation. arXiv preprint arXiv:1906.11455 (2019)

14. Junczys-Dowmunt, M., et al.: Marian: fast neural machine translation in C++. In: Proceedings of ACL 2018, System Demonstrations, pp. 116–121, July 2018

15. Hu, B., Han, A., Zhang, Z., Huang, S., Ju, Q.: Tencent minority-mandarin translation system. In: Huang, S., Knight, K. (eds.) CCMT 2019. CCIS, vol. 1104, pp. 93–104. Springer, Singapore (2019). https://doi.org/10.1007/978-981-15-1721-1_10

16. Johnson, M., et al.: Google's multilingual neural machine translation system: enabling zero-shot translation. Trans. Assoc. Comput. Linguist. (TACL) 5, 339–351 (2017)

17. Dyer, C., Chahuneau, V., Smith, N.A.: A simple, fast, and effective reparameterization of IBM model 2. In: Proceedings of the 2013 Conference of the North American Chapter of the Association for Computational Linguistics: Human Language Technologies (NAACL-HLT 2013), pp. 644–648, June 2013

18. Yee, K., Dauphin, Y., Auli, M.: Simple and effective noisy channel modeling for neural machine translation. In: Proceedings of the 2019 Conference on Empirical Methods in Natural Language Processing and the 9th International Joint Conference on Natural Language Processing (EMNLP-IJCNLP 2019), pp. 5700–5705, November 2019

19. Biçici, E., Yuret, D.: Instance selection for machine translation using feature decay algorithms. In: Proceedings of the Sixth Workshop on Statistical Machine Translation (WMT 2011), pp. 272–283, July 2011

20. Post, M.: A call for clarity in reporting BLEU scores. In: Proceedings of the Third Conference on Machine Translation: Research Papers (WMT 2018), pp. 186–191, October 2018

21. Huang, C., Zhao, H.: Chinese word segmentation: a decade review. J. Chin. Inf. Process. 21(3), 8–19 (2007)

22. Zhao, H., Cai, D., Huang, C., Kit, C.: Chinese word segmentation: another decade review (2007–2017). arXiv preprint arXiv:1901.06079 (2019)

Tsinghua University Neural Machine Translation Systems for CCMT 2020

Gang Chen[1], Shuo Wang[1], Xuancheng Huang[1], Zhixing Tan[1], Maosong Sun[1,2], and Yang Liu[1,2(✉)]

[1] Institute for Artificial Intelligence,
Department of Computer Science and Technology,
Tsinghua University, Beijing, China
liuyang2011@tsinghua.edu.cn
[2] Beijing National Research Center for Information Science and Technology,
Beijing Academy of Artificial Intelligence, Beijing, China

Abstract. This paper describes the neural machine translation system of Tsinghua University for the bilingual translation task of CCMT 2020. We participated in the Chinese ↔ English translation tasks. Our systems are based on Transformer architectures and we verified that deepening the encoder can achieve better results. All models are trained in a distributed way. We employed several data augmentation methods, including knowledge distillation, back-translation, and domain adaptation, which are all shown to be effective to improve translation quality. Distinguishing original text from translationese can lead to better results when performing domain adaptation. We found model ensemble and transductive ensemble learning can further improve the translation performance over the individual model. In both Chinese → English and English → Chinese translation tasks, our systems achieved the highest case-sensitive BLEU score among all submissions.

1 Introduction

This paper describes the neural machine translation (NMT) systems of Tsinghua University for the CCMT 2020 translation task. We participated in two directions of bilingual translation tasks: Chinese → English and English → Chinese. We exploited the following techniques to build our systems:

- Deep Transformers: We train deep transformer models with mixed-precision and distributed training.
- Data augmentation: We explored various data augmentation methods such as back-translation and knowledge distillation.
- Finetuning and model ensemble: We use finetuning and model ensemble to further improve the performance of our systems.

G. Chen, S. Wang and X. Huang—Equal contributions. Listing order is random.

© Springer Nature Singapore Pte Ltd. 2020
J. Li and A. Way (Eds.): CCMT 2020, CCIS 1328, pp. 98–104, 2020.
https://doi.org/10.1007/978-981-33-6162-1_9

The overview of our methods is shown in Fig. 1. The remainder of this paper is structured as follows: Sect. 2 describes the methods used in our CCMT 2020 submissions. Section 3 shows the settings and results of our experiments. Finally, we conclude in Sect. 4.

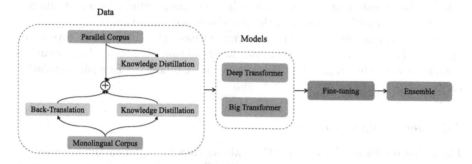

Fig. 1. Overview of Tsinghua NMT systems.

2 Methods

2.1 Data

We use all available data provided by CCMT and WMT, which contains a total of 26.7M bilingual sentence pairs. We apply the following procedures to preprocess the data:

- We remove illegal UTF-8 characters and replace all control characters with a space character.
- All Traditional Chinese sentences are converted into Simplified Chinese ones.
- We apply Unicode NFKC normalization to normalize texts.
- We further restore HTML/XML escape and normalize punctuation in the texts.

The resulting parallel corpus still contains many noise sentence pairs. Therefore, we further filter the data using the following rules:

- We remove all duplicate sentence pairs in the data.
- All sentences that contain illegal characters (e.g., Chinese characters in English sentences) are discarded.
- We translate both the Chinese and English sides of the bilingual data with baseline NMT systems and compute the sentence-level BLEU scores between the translated and original sentences. Then we discard all sentence pairs with BLEU scores below 5.

After filtering, the final data used in our experiments contains about 21M sentence pairs. To tokenize the texts, we use Jieba segmenter[1] for Chinese and Moses toolkit[2] for English.

---

[1] https://github.com/fxsjy/jieba.
[2] https://github.com/moses-smt/mosesdecoder.

2.2   Models

Deep Transformer. According to the previous work [6], the performance of Transformer models [7] can be improved by increasing the number of layers in the encoder. We follow [6] to use deep Transformer in our experiments. To address the vanishing-gradient problem in deep Transformer, we adopt the pre-layer normalization [8] instead of the post-layer normalization [7].

In our experiments, both the big Transformer with 15 encoder layers and the base Transformer with 50 encoder layers obtain significant improvements compared with the vanilla big Transformer on Chinese → English translation task.

2.3   Data Augmentation

Back Translation. We augment the training data by exploring the monolingual corpus using back translation [1,4]. Specifically, we select a portion of sentences in the target monolingual corpus and then translate them back into the source language using target-to-source (T2S) models. We merge the synthetic data with the bilingual data to train our models. Following Edunov et al. [1], we also add noise to the translated sentences to further improve the performance.

Knowledge Distillation. We further augment the data by applying sequence-level knowledge distillation (KD) [2]. We explore the following types of KD in our experiments:

- R2L KD: We replace the target-side sentences of the parallel corpus with sentences translated by Right-to-Left (R2L) models.
- Ensemble KD: We ensemble multiple models to translate the source-side sentences in the parallel corpus.
- Monolingual KD: We exploit additional source-side monolingual data by translating them using existing NMT models.

2.4   Finetuning

Previous work [6] found that finetuning with in-domain data can bring huge improvements. We also use development sets as the in-domain datasets. As mentioned in Sun et al. [6], the source side of newsdev2017, newstest2017 and newstest2018 are composed of two parts: documents created originally in Chinese and documents created originally in English. We split these datasets into original Chinese part and original English part according to tag attributes of SGM files. For Chinese-English translation, we use CWMT2008, CWMT2009 and original Chinese part of newsdev2017, newstest2017 and newstest2018 as the in-domain dataset. For English-Chinese translation, we use original English part of newsdev2017, newstest2017 and newstest2018 as the in-domain dataset.

During finetuning, we use a larger dropout rate and a smaller constant learning rate than those in the training process. The model parameters are updated after each epoch, which is enabled by gradient accumulation. We finetune all models with 18 epochs.

2.5   Ensemble

Model ensemble is a well-known technique to combine different models for stronger performance. We utilize the frequently used method for ensemble, which calculates the word-level averaged log-probability among different models during decoding. On account of that the model diversity among single models has a strong impact on the performance of ensemble model, we combine single models that have different model architectures (e.g., different number of encoder layers or different widths of the feed-forward layer) and been trained on different data (e.g., generated by different data augmentation method).

We also try to use Transductive Ensemble Learning (TEL) [9] instead of standard ensemble. TEL is a technique utilizing the synthetic test data (consists of original source sentences and translations of target-side sentences) of different models to finetune a single model. In our experiments, we find that once several single models have been applied TEL, their ensemble model could not outperform single models. We employ 5 left-to-right models and 2 right-to-left models to generate synthetic test data and finetune our best single model. Finally, we get a single model which even outperforms the ensemble of several single models.

Table 1. BLEU evaluation results on the `newstest2019` Chinese-English test set.

Settings	Transformer big	Deep transformer base	Deep transformer big
Baseline	27.94	–	–
+Data augmentation	28.59	29.74	29.85
+Finetuning	35.97	37.48	37.74
Ensemble	**38.95**		

3   Experiments

3.1   Settings

We use the PyTorch implementation of open-source toolkit THUMT[3] to carry out all experiments. To enable open vocabulary, we learn 32K BPE [5] operations separately on Chinese and English texts using `subword-nmt`[4] toolkit. All models are trained on 2 machines with 10 RTX 2080Ti GPUs on each machine.

For all our models, we adopt Adam [3] ($\beta_1 = 0.9, \beta_2 = 0.999$, and $\epsilon = 1 \times 10^{-8}$) as the optimizer. We use the default hyperparameters provided by

---

[3] https://github.com/THUNLP-MT/THUMT.
[4] https://github.com/rsennrich/subword-nmt.

THUMT to train Transformer models. In addition, we use distributed training and mixed-precision training to reduce the training time. Specifically, we set the batch size to 2048 source and target tokens on each GPU per step, and accumulate the gradients for 10 steps to update the model parameters. We train each model for only 50k steps. It takes about 2 days to train a deep Transformer model. After training, we average 5 top checkpoints validated on the validation set [10] and then perform finetuning on top of this new checkpoint.

Table 2. Detailed experiments on different data augmentation methods. All BLEU scores are reported after finetuning on development sets.

Dataset	Baseline	+BT	+Noise BT	+BT&EKD	+BT&R2LKD
newstest2019	34.85	34.53	35.05	35.65	35.02

3.2 Results on Chinese-English Translation

Table 1 shows the results of Chinese-English Translation on newstest2019 dataset. All methods we used can bring substantial improvements over the baseline system. Applying data augmentation methods improves the baseline system by 0.65 BLEU score. The deep models can further bring 1.26 BLEU improvements. In our experiments, finetuning with in-domain data is the most effective approach, which gains about 7–8 BLEU improvements. Furthermore, the gap between Transformer Big and Deep Transformer Big model enlarges after applying the finetuning step.

Table 2 shows the detailed results of different data augmentation methods. The results are reported after applying the finetuning step to see whether the methods can bring further improvements. We have the following findings:

- Back translation does not work well on Chinese-English translation. Considering finetuning on texts translated from Chinese is very effective, we conjecture that the result is caused by the mismatch of style between texts originally from English and texts translated from Chinese to English.
- Adding noise to pseudo-source sentences is helpful to improve translation quality.
- Knowledge distillation methods, such as Ensemble KD and R2L KD, are effective in Chinese-English translation.

As a result, we use all data augmentation methods described above to train our final models. After ensemble 5 deep models, we obtain 11.01 BLEU improvements over the baseline system. Our final submission with TEL achieves 48.12 BLEU-SBP on the ccmt2020 test set, which gains 1.1 BLEU-SBP improvements over the submission with standard model ensemble.

Table 3. BLEU evaluation results on the `newstest2019` English-Chinese test set.

Dataset	Data augmentation		Deep model	Finetuning	Ensemble
	BT	Noise BT			
`newstest2019`	36.94	37.02	36.91	38.33	39.56

3.3 Results on English-Chinese Translation

Table 3 shows the results of English-Chinese Translation on `newstest2019` test set. Due to limited resources, we only use the back-translation method to augment data in this task. As we can see from the table, the results of BT and noise BT are nearly identical, which do not coincides with the findings in Chinese-English translation. Furthermore, we do not found the benefits of deep models on this task. Finetuning with in-domain data brings substantial improvements, but not as effective as Chinese-English translation. After ensembling 4 models finetuned with in-domain data, we finally obtain 39.56 BLEU on `newstest2019`. Our final submission achieves 63.43 BLEU-SBP on `ccmt2020` English-Chinese test set.

4 Conclusion

This paper described the neural machine translation systems developed by Tsinghua University in the CCMT 2020 bilingual translation tasks. We exploited deep models, various data augmentation methods, finetuning techniques, as well as model ensembles in our experiments. We verified through experiments that combining all these methods can lead to substantial improvements in translation quality.

Acknowledgements. This work was supported by the National Key R&D Program of China (No. 2017YFB0 202204), National Natural Science Foundation of China (No. 61925601, No. 61761166 008, No. 61772302), Beijing Academy of Artificial Intelligence, and the NExT++ project supported by the National Research Foundation, Prime Ministers Office, Singapore under its IRC@Singapore Funding Initiative.

References

1. Edunov, S., Ott, M., Auli, M., Grangier, D.: Understanding back-translation at scale. In: Proceedings of EMNLP (2018)
2. Kim, Y., Rush, A.M.: Sequence-level knowledge distillation. In: Proceedings of EMNLP (2016)
3. Kingma, D.P., Ba, J.: Adam: a method for stochastic optimization. In: Proceedings of ICLR (2015)
4. Sennrich, R., Haddow, B., Birch, A.: Improving neural machine translation models with monolingual data. In: Proceedings of ACL (2016)

5. Sennrich, R., Haddow, B., Birch, A.: Neural machine translation of rare words with subword units. In: Proceedings of ACL (2016)
6. Sun, M., Jiang, B., Xiong, H., He, Z., Wu, H., Wang, H.: Baidu neural machine translation systems for WMT19. In: Proceedings of the Fourth Conference on Machine Translation (2019)
7. Vaswani, A., et al.: Attention is all you need. In: Proceedings of NeurIPS (2017)
8. Wang, Q., et al.: Learning deep transformer models for machine translation. In: Proceedings of ACL (2019)
9. Wang, Y., Wu, L., Xia, Y., Qin, T., Zhai, C., Liu, T.Y.: Transductive ensemble learning for neural machine translation. In: AAAI (2020)
10. Wei, H.R., Huang, S., Wang, R., Dai, X., Chen, J.: Online distilling from checkpoints for neural machine translation. In: Proceedings of NAACL-HLT (2019)

BJTU's Submission to CCMT 2020 Quality Estimation Task

Hui Huang, Jin'an Xu[✉], Wenjing Zhu, Yufeng Chen, and Rui Dang

School of Computer and Information Technology, Beijing Jiaotong University,
Beijing, China
{18112023, jaxu, 18120461, chenyf, 19125167}@bjtu.edu.cn

Abstract. This paper presents the systems developed by Beijing Jiaotong University for the CCMT 2020 quality estimation task. In this paper, we propose an effective method to utilize pretrained language models to improve the performance of QE. Our model combines three popular pretrained models, which are Bert, XLM and XLM-R, to create a very strong baseline for both sentence-level and word-level QE. We tried different strategies, including further pretraining for bilingual input, multi-task learning for multi-granularities and weighted loss for unbalanced word labels. To generate more accurate prediction, we performed model ensemble for both granularities. Experiment results show high accuracy on both directions, and outperform the winning system of last year on sentence level, demonstrating the effectiveness of our proposed method.

Keywords: Machine Translation · Quality Estimation · Pretrained language model

1 Introduction

Machine translation quality estimation (Quality Estimation, QE) aims to evaluate the quality of machine translation automatically without golden reference. QE can be implemented on different granularities, thus to give an estimation for different aspects of machines translation output.

This paper introduces in detail the submission of Beijing Jiaotong University to the quality estimation task in the 16th China Conference on Machine Translation (CCMT2020). This year, the QE task consists of two language directions of Chinese-English and English-Chinese, and two granularities of word-level and sentence-level subtasks, thus four subtasks in total.

We propose an effective method to utilize pretrained language models to improve the performance of QE. Our model combines three popular pretrained models, which are Bert [1], XLM [2] and XLM-R [3], to create a very strong baseline for both sentence-level and word-level QE.

We also tried different strategies to boost the final results, including further pretraining for bilingual input, multi-task learning for multi-granularities and weighted loss for unbalanced word labels. To improve the final accuracy, we ensembled the results generated by different models for both sentence and word level.

J. Li and A. Way (Eds.): CCMT 2020, CCIS 1328, pp. 105–113, 2020.
https://doi.org/10.1007/978-981-33-6162-1_10

Experiment results show that our model achieves high accuracy on both directions, surpassing previous models on sentence-level, and obtaining competitive performance on word-level, demonstrating the effectiveness of our proposed method.

2   Model Description

2.1   Pretrained Models for Quality Estimation

Our method is based on three recent proposed pretrained models, Bert, XLM and XLM-R. All of these three models are based on multi-layer Transformer architecture with different training procedures.

For both word-level and sentence-level QE task, we concatenate source sentences and machine translated sentences following the way pre-trained models treat sentence pairs, and do prediction on the top of them.

For sentence-level prediction, we tried two different strategies. The first one is to directly use the first token according to the special token [CLS] to perform prediction, as we believe this logit integrates sentence-level information. The second one is to add another layer of RNN on the top of pre-trained models, to better leverage the global context information, as shown in Fig. 1.

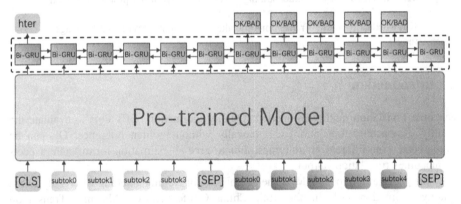

Fig. 1. Pre-trained model for QE with multi-task learning. The component circled with dashed line is alternative.

For word-level prediction, we use each logit according to each token in the sentence to generate word-quality label.

The loss functions for word and sentence-level are as follows:

$$L_{word} = \sum_{s \in D} \sum_{x \in s} -(p_{OK} \log p_{OK} + p_{BAD} \log p_{BAD}) \tag{1}$$

$$L_{sent} = \sum_{s \in D} \|sigmoid(W_s h(s)) - hter_s\| \tag{2}$$

where s and x denote each sentence and word in the dataset, p_{OK} and p_{BAD} denote the probability for each word to be classified as OK/BAD, $h(s)$ denotes the hidden representation for each sentence, and W_s denote the transformation matrices for sentence and word level prediction, and $hter_s$ denote the HTER[1] measure for each sentence.

2.2 Further Pretraining for Bilingual Input

Despite the shared multilingual vocabulary, Bert is originally a monolingual model, treating the input as either being from one language or another. To help Bert adapts to sentence pairs from different languages, we implement a further pretraining step, training Bert model with massive parallel machine translation data [4].

For our task of QE, we combine bilingual sentence pairs from large-scale parallel dataset, and randomly mask sub-word units with a special token, and then train Bert model to predict masked tokens. Since our input are two parallel sentences, during the predicting of masked words given its context and translation reference, Bert can capture the lexical alignment and semantic relevance between two languages.

After this further pretraining step, Bert model is familiar with bilingual inputs, and acquires the ability to capture translation errors between different languages. This method is similar to the pretraining strategy mask-language-model in [1], while its original implementation is based on only sentences from monolingual data.

In contrast, XLM and XLM-R are multilingual models which receive two sentences from different languages as input, which means further pretraining is likely to be redundant. This is verified by our experiment results demonstrated in the following section.

2.3 Multi-task Learning for Multi-granularities

The QE subtasks at sentence and word-level are highly related because their quality annotations are commonly based on the HTER measure. Quality annotated data of other subtasks could be helpful in training a QE model specific to a target task [5].

We also implemented multi-task learning on our pretrained models. Since the linear transformation for predictions according to different granularities are implemented on different positions, we can perform multi-task training and inference naturally without any structure adjustment. Since we tried two different models, with or without bidirectional RNN, our model can be illustrated as the following figure:

During training, predictions for different granularities are generated at the same time on different positions, and losses are combined and back-propagated simultaneously. The loss function is as follows:

$$L_{join} = \sum_{s \in D} \sum_{x \in s} cross\_entropy(W_w h(x), y_x) + \|sigmoid(W_s h(s)) - hter_s\| \quad (3)$$

where $h(x)$ and $h(s)$ denote the hidden representations for each word and sentence, and W_w and W_s denote the transformation matrices for sentence and word prediction.

---

[1] https://github.com/jhclark/tercom.

Most model components are common across sentence-level and word-level tasks except for the output matrices of each task, which is especially beneficial for sentence-level prediction, since the training objective for sentence QE only consists of one single logit containing limited information.

2.4 Weighted Loss for Unbalanced Word Labels

The quality of machine translated sentences in QE data is very high [6], which means that a huge proportion of the sentences do not need post-editing at all. This leads to an unbalanced label distribution where most of the word labels are BAD, which makes it very likely to give a skewed prediction with a very low F1 score for BAD words.

To improve the overall performance, we add up to the weight for BAD words when calculating cross-entropy loss, enabling the model emphasize more on the incorrectly translated words. The word-level loss function is as follows:

$$L_{word} = \sum_{s \in D} \sum_{x \in s} -(p_{OK} \log p_{OK} + \lambda p_{BAD} \log p_{BAD}) \tag{4}$$

where λ is a hyper-parameter larger than 1.

We also tried other data augmentation skills to balance word labels, which is demonstrated in the next section.

2.5 Multi-model Ensemble

Until now, we have built three different QE models trained with different architectures, which can capture different information from the same text. Considering the variation of different strategies and initialized parameters, we can have multiple models for each subtask, which can be integrated to generate stronger performance [7].

For word level QE, to ensemble multiple predictions for each token, we tried two different strategies. The first one is to take the average of logit generated by softmax layer for each token, and then argmax it to get OK/BAD label. The second one is to vote based on different labels generated by different models. For an instance, if there are two Oks and one BAD out of three predictions for a token, then the ensembled result for this token would be OK.

For sentence level QE, we simply take the average of predicted HTER scores from different models.

Due to time limitation, we did not explore more complex ensemble techniques illustrated in [8], which introduced conspicuous improvement in their work.

3   Experiment

3.1   Dataset

We use the QE data from CCMT2020 Machine Translation Quality Estimation tasks. CCMT QE tasks contain two different language directions (Chinese-English and English-Chinese) on both sentence-level and word-level. The amount of data provided on both language pairs and levels are very small (no more than 15 k triples on all directions), which makes QE a highly data-sparse task.

To further pretrain the Bert model, we use the parallel dataset for Chinese-English Translation task in CCMT2020, which contains nearly 7 million sentence pairs.

3.2 Experiment Results

The experiment results on both directions and granularities are shown in Table 1 and Table 2, where *transformer-dlcl* [9] and *CCNN* were the top2 systems in CCMT 2019 QE task.

For sentence-level QE, we surpass all baselines on both directions with limited computation resource. For word-level QE, we do not manage to surpass the top 1 system of last year. But we have to mention that on word-level task, we do not apply further pretraining step on both models before finetuning, so the computation overhead is very low with just a few hours fine-tuning on one single GPU.

Table 1. Experiment results on CCMT2020 sentence-level QE dev set

Language Direction	System	Pearonr	Spearman	MSE
Chinese- English	CCNN	0.56	0.49	–
	transformer-dlcl	0.6164	–	–
	Bert	0.6069	0.5182	0.5626
	XLM	0.5744	0.5467	0.5606
	XLM-R	0.5657	0.5057	0.5357
	Ensemble Model	**0.6277**	0.5701	–
English-Chinese	CCNN	0.55	0.42	–
	transformer-dlcl	0.5861	–	–
	Bert	0.5172	0.3907	0.4540
	XLM	0.5540	0,4110	0.4825
	XLM-R	0.5365	0.4001	0.4683
	Ensemble Model	**0.5907**	0.5521	–

Table 2. Experiment results on CCMT2020 word-level QE dev set

Language Direction	System	F1-Multi	F1-BAD	F1-OK
Chinese- English	transformer-dlcl	**0.5393**	0.6152	0.8767
	Bert	0.4846	0.5634	0.8602
	XLM	0.4844	0.5635	0.8597
	XLM-R	0.5061	0.5902	0.8575
	Ensemble Model	0.5141	0.5913	0.8649
English-Chinese	transformer-dlcl	**0.4385**	0.8974	0.4886
	Bert	0.3947	0.4508	0.8757
	XLM	0.4073	0.4625	0.8808
	XLM-R	0.4173	0.4669	0.8973
	Ensemble Model	0.4336	0.4841	0.8958

Besides, we do not introduce any pseudo data during the training of our QE system, while transformer-dlcl introduced pseudo data in all subtasks, which led to the improvement of 2-4 points.

In a word, the pretrained language model can be a very strong baseline for QE at both sentence-level and word-level. It requires no complicated architecture engineering and massive training data, and can provide reliable performance.

3.3   Ablation Study

In this section, we will discuss the influence of different strategies on our model. Notice although we described a lot of strategies to boost QE system in former sections, their influence on different granularities are different. Besides, due to the update of codes during the evaluation period, there may be some discrepancy between the following results and the results we released in Sect. 3.2.

Extra Bi-RNN. It is alternative to add an extra layer of bidirectional RNN before the softmax layer. Extra layer may introduce more globalized prediction, but may also introduce noise since we have to random-initialize it.

As shown in Table 3, an Extra layer of Bi-RNN does not necessarily introduce improvement. Sometimes it can and sometime it cannot. But If there is no multi-task learning when doing sentence-level QE, then an extra layer is compulsory for XLM and XLM-R, since these two models are not pretrained with sentence-level task.

Further Pre-training for Bilingual Input. As we have mentioned before, Bert is only trained with monolingual input, so it is reasonable to believe further pre-training could help Bert adapted to multilingual input. But astonishingly, we find further pre-training can only improve the sentence-level QE, and is harmful for word-level QE on Bert, as shown in Table 4, which needs our future investigation.

Table 3. Extra Bi-RNN on the top of pre-trained model

Language Direction	System	Level	Extra Bi-RNN	F1-multi
Chinese-English	XLM-R	sentence	No	0.5386
			Yes	0.5657
		word	No	0.5057
			Yes	0.4993
	XLM	sentence	No (w/o muti-task)	0.0975
			No	0.5744
			Yes	0.5666

Multi-task Learning for Multi-granularities. As shown in Table 5, after joint trained with different granularities, the results of sentence-level QE increase a lot, which verifies our conjecture that word-level labels can help the training of sentence-level QE. For word-level QE, the avail of multi-task learning seems limited.

Table 4. Further pre-training for bilingual input

Language direction	System	Level	Further pretrain	Pearsonr/F1-multi
English-Chinese	Bert	sentence	No	0.4230
			Yes	0.5169
		word	No	0.3902
			Yes	0.3837

Label Balancing for Word-level QE. We try three different strategies including up-sampling sentence-pairs with high HTER values and down-sampling sentence-pairs with low HTER values, and find that weight balancing when calculating loss is a simple yet effective strategy, as shown in Table 6.

Table 5. Multi-task learning for multi-granularities

Language direction	Level	Model	Multi-task	Pearsonr/F1-multi
English-Chinese	sentence	Bert	No	0.4893
			Yes	0.5169
	word	Bert	No	0.3962
			Yes	0.3902

Although data sampling can also help the model to emphasize more on the bad words when training, but it will damage the natural distribution of sentence-pairs, and thus harmful to final performance. We try different values for λ ranging from 5 to 20, and finally set λ as 10 in Eq. (4).

Table 6. Label balancing for word QE

Language direction	Level	Model	Balancing strategy	F1-multi
English-Chinese	word	Bert	No	0.3227
			up sampling	0.3847
			down sampling	0.3357
			weight balancing	0.3962

Word-level Multi-model Ensemble. As we have mentioned before, there are two strategies to do word-level ensemble, namely averaging logits and voting. Intuitively, averaging logits should be more effective, since more information is integrated. But experiment defies our hypothesis, as show in Table 7.

As shown in Table 7, we did not see significant outperformance of logit averaging over voting mechanism. This may be caused by the unbalanced word-label, which leading to a biased logit distribution (where most tokens are assigned with a logit close to 1). Even there is one prediction under 0.5, it would not change the result since the other predictions are likely to be almost 1 produced by the softmax layer.

Table 7. Word-level Multi-model Ensemble

Language direction	Level	Model	Balancing strategy	F1-multi
English-Chinese	word	Ensembled	voting	0.4321
			logit averaging	0.4336
Chinese-English	word	Ensembled	voting	0.5116
			logit averaging	0.5141

4   Conclusion

In this paper, we described our submission in quality estimation task, consisting of two language directions and two granularities. We implement the QE system based on three popular pretrained models, namely Bert, XLM and XLM-R, and study different applicable strategies on QE task, i.e. further pretraining on bilingual input, multi-task training on multi-granularities and weighted loss for word labels. We ensembled multiple models to generate more accurate prediction. Our model achieves strong performance on both sentence-level and word-level QE tasks with limited computation resource, and outperforms the previous SOTA models on sentence-level development set, verifying the validity of our proposed strategies.

Massive linguistic knowledge contained in pretrained models is very helpful for the QE task even when there is limited training data. In the future, we will continue our research on the application of pretrained models on different QE tasks.

Acknowledgement. This work is supported by the National Natural Science Foundation of China (Contract 61976015, 61976016, 61876198 and 61370130), and the Beijing Municipal Natural Science Foundation (Contract 4172047), and the International Science and Technology Cooperation Program of the Ministry of Science and Technology (K11F100010).

References

1. Devlin, J., Chang, M.W., Lee, K., Toutanova, K.: Bert: Pre-training of deep bidirectional transformers for language understanding. arXiv preprint arXiv:1810.04805 (2018)
2. Lample, G., Conneau, A.: Cross-lingual language model pretraining. arXiv preprint arXiv:1901.07291 (2019)
3. Conneau, A., et al.: Unsupervised cross-lingual representation learning at scale. arXiv preprint arXiv:1911.02116 (2019)
4. Kim, H., Lim, J.H., Kim, H.K., Na, S.H.: QE BERT: bilingual BERT using multi-task learning for neural quality estimation. In: Proceedings of the Fourth Conference on Machine Translation (Volume 3: Shared Task Papers, Day 2) (2019)

5. Hyun, K., Jong-Hyeok, L., Seung-Hoon, N.: Predictor-estimator using multilevel task learning with stack propagation for neural quality estimation. In: Proceedings of the Second Conference on Machine Translation, Volume 2: Shared Task Papers, pp. 562–568 (2017)

6. Specia, L., Blain, F., Logacheva, V., Astudillo, R., Martins, A.F.: Findings of the wmt 2018 shared task on quality estimation. In: Proceedings of the Third Conference on Machine Translation: Shared Task Papers, pp. 689–709 (2018)

7. Zhi-Hua, Z., Jianxin, W., Wei, T.: Ensembling neural networks: many could be better than all. Artif. Intell. **137**(1–2), 239–263 (2002)

8. Kepler, F., et al.: Unbabel's Participation in the WMT19 Translation Quality Estimation Shared Task. arXiv preprint arXiv:1907.10352 (2019)

9. Wang, Z., et al.: NiuTrans Submission for CCMT19 Quality Estimation Task. In: Huang, S., Knight, K. (eds.) CCMT 2019. CCIS, vol. 1104, pp. 82–92. Springer, Singapore (2019). https://doi.org/10.1007/978-981-15-1721-1_9

NJUNLP's Submission for CCMT20 Quality Estimation Task

Qu Cui, Xiang Geng, Shujian Huang$^{(\boxtimes)}$, and Jiajun Chen

National Key Laboratory for Novel Software Technology, Nanjing University,
Nanjing, China
{cuiq,gx}@smail.nju.edu.cn,
{huangsj,chenjj}@nju.edu.cn

Abstract. Quality Estimation is a task to predict the quality of translations without relying on any references. QE systems are based on neural
features but suffer from the limited size of QE data. The best models
nowadays transfer bilingual knowledge from parallel data to QE tasks.
However, the distribution between parallel data and QE data may lead
to the value of parallel data not being used for best. More specifically,
there are no errors in parallel translations while there may be more than
one error in the translations of QE data. To alleviate this problem, we
propose a model that will mask some tokens at the target side on parallel data but still need to predict every target token. And based on this
model, we propose a variant model that uses a masked language model
at the target side to obtain deep bi-directional information. Besides, we
also try different ensemble methods to get better performance of the
CCMT20 Quality Estimation Task. Our system finally won second place
in the ZH-EN language pair and third place in the EN-ZH language pair.

Keywords: Quality Estimation · Data distribution · Mask

1 Introduction

Quality Estimation is a task to predict the quality of translations without relying
on any references. It has both a word-level and a sentence-level task; all the
quality scores are computed automatically by TERCOM [13].

Researchers first use some hand-craft features to represent the translation
pairs and do QE tasks [7]. However, it is time-consuming and expensive. Then,
automatic neural features are introduced to QE tasks [1,12]. The remaining
problem is that the data of QE tasks is hard to get, and it limits the performance
of QE models. To solve this problem, researchers began to transfer bilingual
knowledge from parallel data to QE tasks. They usually follow a predictor-
estimator framework [3,6], which first trains a predictor on parallel data and
then trains an estimator on QE data based on the features produced by the
predictor.

This structure has achieved great success, but it still has some problems. The
data distribution between QE data and parallel data is different. The translations

© Springer Nature Singapore Pte Ltd. 2020
J. Li and A. Way (Eds.): CCMT 2020, CCIS 1328, pp. 114–122, 2020.
https://doi.org/10.1007/978-981-33-6162-1_11

in QE data are generated by an accurate machine translation (MT) system, and there will inevitably be some noise. However, there are nearly no errors in parallel translations. When training the predictor on parallel data, the model may only make the right choice based only on the translation. In this case, there will be problems when doing QE tasks since the real translations in QE data are no longer reliable.

To alleviate this problem, we propose a model that will mask some tokens at the target side but still need to correctly predict every token. Such a way will help the model reduce the dependence on the translations when training on the parallel data, and it can enhance the ability of the model to deal with translations with errors. Moreover, to obtain the deep bi-directional knowledge, we use a masked language model at the target side instead of a concatenation of two single directional decoders, which is used in the traditional predictor-estimator framework.

We ensembled the existing methods and our proposed methods and participated in the CCMT20 Quality Estimation task. Our system finally won second place in the ZH-EN language pair and the third place in the EN-ZH language pair. Meanwhile, we also conduct experiments to show how our approach works.

2 Methods

In this section, we are going to show the details of the methods used in our final submitted system. They will be divided into two parts. First, we will list the existing methods and second, the proposed methods by us.

2.1 Existing Methods

QUETCH As the name implies, QUETCH [8], QUality Estimation from scra-TCH, is trained from scratch with only QE data. The architecture of QUETCH consists of one embeddings layer, one linear layer with the tanh activation function, one output layer with the softmax activation function. We use the fraction of BAD labels as an estimate for the HTER score at sentence-level [10].

NuQE NuQE [9], NeUral Quality Estimation, carries QUETCH one step further with complex neural networks. Their model architecture consists of a linear layer, a bi-directional GRU layer, two other linear layers. And NuQE is also trained without auxiliary parallel data.

We use QUETCH and NuQE as implemented in OpenKiwi [5][1].

QE Brain QE Brain [3] follows the predictor-estimator architecture. When training the predictor on parallel data, they first use an encoder based on transformer [14] to encode the source sentence $\mathbf{X} = \{x_1, x_2, \ldots, x_n\}$ and

---

[1] https://unbabel.github.io/OpenKiwi.

then use a bi-directional decoder to predict each token in the target sentence $\mathbf{Y} = \{y_1, y_2, \ldots, y_c\}$ with the help of hidden representations of the source sentence.

When training on real QE data, they also use the predictor to predict each token in the translation from real translation systems. And the hidden state of the final layer in the predictor will be used as the word-level features. Meanwhile, the probability difference between the probability of generating the current token and the most likely token, which is called mis-matching feature will also be used as the word-level feature. Finally, they use a Bi-LSTM [4] as the estimator to combine the word-level features to predict the word-level tags $\mathbf{O}$ and sentence-level scores q.

Our proposed models are mainly based on the QE Brain.

2.2   Proposed Methods

Masked QE Brain. Researches used to transfer bilingual knowledge from parallel data to QE tasks, however, the data distribution between parallel data and QE data is different. The main difference is that, a real machine translation system generates the translations in QE data, and there will be some errors. While humans generate the translations in parallel data, and there are nearly no errors. The predictor trained on parallel data can not perform well when it is feeding with translations with errors because the contexts at the target side are different. To partially alleviate this problem, we proposed our Masked QE Brain.

The motivation for our method is very direct. We want to enhance the ability of the model when making predictions with wrong contexts. In order to achieve this goal, we mask some tokens in the translation when training the predictor on parallel data. And the predictor needs to make the same prediction as they are feeding with the complete pair. The rest of our model are the same as those in the original version of the QE Brain.

Masked Target Language Model. The QE Brain and Masked QE Brain use a bi-directional decoder at the target side to obtain the information from both sides. However, this architecture is just a shallow concatenation which can not truly get the bi-directional information [2]. We use a masked language model [2] at the target side instead and get the deep bi-directional information. We call this model the Masked Target Language Model (MTLM), and the format of the input is just the same as that in the Masked QE Brain. The two models use the source sentence $\mathbf{X}$, the masked target sentence $\mathbf{Y}'$ as the input. And the MTLM only need to predict the right tokens of these masked ones at the target side while Masked QE Brain needs to predict all the tokens.

Figure 1 shows the model architecture of the original QE Brain and the two proposed models.

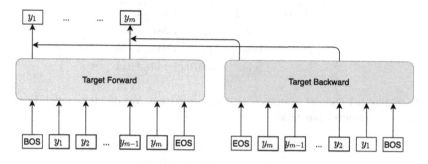

(a) Original QE Brain.

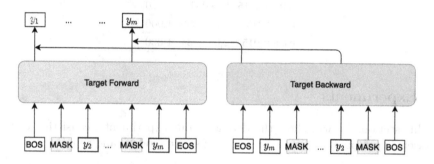

(b) Masked QE Brain.

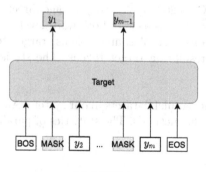

(c) MTLM.

Fig. 1. These models have the same source encoder; we do not show it in the figure to save space. (a) shows the original QE Brain, and (b) enhances it by simply masking tokens at the target side. (c) uses a masked language model at the target side to obtain deep bidirectional information.

Table 1. QE Dataset statistics of the CCMT20.

Direction	Aspect	Train	Dev	Test
EN-ZH	Word	10,878	1,128	4,151
	Sent	14,789	1,381	4,355
ZH-EN	Word	11,017	1,046	4,129
	Sent	10,070	1,143	4,211

Table 2. Parallel Dataset statistics used in our system. We divide parallel data into training set and development set.

Dataset	Train	Dev
WMT18	7,460,939	2,000
neu2017	1,999,000	1,000
datum2015	999,004	1,000
casia2015	1,049,000	1,000
casict2015	2,035,834	1,000

3   Experiments

In this section, we will show the details of our experiments, consisting of the dataset, hyper-parameters, performance of single models, and so on.

3.1   Dataset

QE Dataset. The QE tasks of CCMT2020 have two language directions of both EN-ZH and ZH-EN, and they have two aspects of both word-level and sentence-level. The word-level task contains tags of source tokens, target tokens, and target gaps, and we only have results on target tokens. The statistics of QE datasets are shown in Table 1.

Parallel Dataset. We use different parallel datasets in our system. And we do not use all parallel datasets on all of the methods. The statistics of parallel datasets are shown in Table 2.

3.2   Settings

Metrics. For the word-level task, the metrics are F1 scores of the products of both positive and negative examples. For the sentence-level task, the metric is Pearson's Correlation Coefficient.

Table 3. Results of the CCMT20.

Parallel Dataset	Method	EN-ZH		ZH-EN	
		Sent	Word	Sent	Word
–	NuQE	19.39	**29.22**	23.75	35.56
–	QUETCH	29.08	11.72	25.04	30.24
WMT18	QE Brain	47.26	15.90	52.04	37.68
	Masked QE Brain	47.26	23.38	52.77	35.60
	MTLM	**52.16**	24.67	**55.94**	**43.75**
neu2017	MTLM	45.23	20.07	53.43	40.62
datum2015		44.58	14.49	50.49	41.30
casia2015		44.27	23.43	51.45	42.34
casict2015		39.19	14.74	52.34	42.53
ensemble	Sent-neural	56.55	–	57.23	–
Ensemble	Sent-result	54.18	–	55.18	–
Ensemble	Word-voting	–	30.25	–	48.28

Hyper-parameters. For NuQE and QUETCH, we simply use the software released publicly.

For the original QE Brain and Masked QE Brain, both the encoders and the two decoders have 6 layers of transformers with 512 hidden units. And for MTLM, the model only has one encoder and one decoder, which has the same number of layers and units as the original QE Brain. These three models all use Bi-LSTM [4] as the estimator, of which the hidden size is set to 512.

Tokenize. We use BPE [11] to tokenize the English dataset and use jieba[2] to tokenize the Chinese dataset. The step of BPE is set to 30,000, and we use all tokens after tokenized. And we use *jieba* to tokenize the Chinese sentences. Meanwhile, the tokens in the EN-ZH word-level task will not be tokenized. Finally, we only use the 30,000 most frequent tokens of all Chinese tokens.

3.3   Single Model Results

The results of single models are shown in the Table 3. As we can see, these models without parallel knowledge do not have a good performance except on the word-level task of the EN-ZH direction. When pretraining on the WMT18 parallel dataset, our two proposed methods all perform better than the original QE Brain. And the MTLM has the best performance due to that it both alleviates the problem of data distribution and obtains deep bi-directional information.

More parallel data will bring better performance. The MTLM's performance pretraining on the data of WMT18 is better than that of the other four datasets, and the size of the WMT18 dataset is almost three times bigger.

---

[2]   https://pypi.org/project/jieba/.

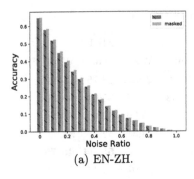
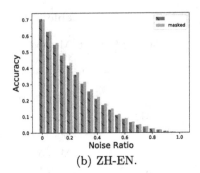

(a) EN-ZH.   (b) ZH-EN.

Fig. 2. The accuracy of predicting right tokens when a part of target tokens are replaced by random ones.

3.4 Ensemble

We try two different ensemble methods at the sentence-level.

Neural Ensemble. The hidden states of QE Brain, Masked QE Brain, MTLM for the same sentence will be gathered, and then an extra linear model will be used to map the hidden states to real HTER values.

Result Ensemble. We gather the HTERs of both training datasets and development datasets from all of the models described above. We then train a linear model that learns to use these HTER values to predict the golden HTER value.

And for the word-level task, we simply use voting to ensemble the results of all models.

The ensemble results are also shown in Table 3. And we can see that the neural ensemble way outperforms the other one at the sentence-level. Our system finally won second place in the ZH-EN language pair and the third place in the EN-ZH language pair.

4   Analysis

In this section, we will discuss the effectiveness of our approach.

Table 4. A case of training data in the parallel data.

SRC	for those toiling far below the surface of the Earth , the proposed system could prove a godsend .
TGT	对于 那些 在 地下 辛苦工作 的 人 来说 ， 这个 新 系统 简直是 神赐 之物 。
Masked TGT	对于 那些 在 地下 [MASK] 的 人 来说 ， 这个 新 系统 简直是 [MASK] [MASK] 。

Table 4 shows an example of our training data in the parallel dataset. As we can see, when training the predictor on the complete target sentence, the model may predict the token '之物' only with the help of token '神赐', because this binary combination is common. However, what we want is that the model can rely less on target sentences. We can easily break the self-dependence by masking tokens in target sentences, and this will enhance the ability of the predictor when feeding with wrong target sentences.

We test the predicting ability of the original QE Brain and our Masked QE Brain when the target sentences are partially replaced by random tokens. The results are shown in Fig. 2. And we can see that when there is no noise in target sentences, the two models have a similar performance. As the replacement ratio grows, our Masked QE Brain has a growing advantage in both language directions.

5 Conclusion

This paper describes our systems for CCMT20 Quality Estimate tasks, including both word-level and sentence-level.

We follow the predictor-estimator architecture and mainly follow QE Brain. To alleviate the problem that the distribution between parallel data and QE data is different, we proposed the Masked QE Brain. And to achieve the deep bi-directional information, we use a masked language model at the target side and propose our MTLM.

The proposed models perform better than the original version of the QE Brain. At the same time, we use different ensemble methods to achieve our final results for CCMT20. Our system finally won second place in the ZH-EN language pair and the third place in the EN-ZH language pair.

Acknowledgement. The authors would like to thank Yiming Yan for the feedback.

References

1. Chen, Z., Tan, Y., Zhang, C., Xiang, Q., Wang, M.: Improving machine translation quality estimation with neural network features. In: Proceedings of the Second Conference on Machine Translation (2017)
2. Devlin, J., Chang, M.W., Lee, K., Toutanova, K.: Bert: pre-training of deep bidirectional transformers for language understanding. In: Proceedings of the 2019 Conference of the North American Chapter of the Association for Computational Linguistics: Human Language Technologies (2018)
3. Fan, K., Wang, J., Li, B., Zhou, F., Chen, B., Si, L.: "bilingual expert" can find translation errors. In: Proceedings of the AAAI Conference on Artificial Intelligence (2018)
4. Graves, A., Schmidhuber, J.: Framewise phoneme classification with bidirectional LSTM and other neural network architectures. Neural Networks (2005)
5. Kepler, F., Trníous, J., Treviso, M., Vera, M., Martins, A.F.T.: Openkiwi: an open source framework for quality estimation (2019)

 6. Kim, H., Lee, J.H., Na, S.H.: Predictor-estimator using multilevel task learning with stack propagation for neural quality estimation. In: Proceedings of the Second Conference on Machine Translation (2017)
 7. Kozlova, A., Shmatova, M., Frolov, A.: YSDA participation in the wmt'16 quality estimation shared task. In: Proceedings of the First Conference on Machine Translation (2016)
 8. Kreutzer, J., Schamoni, S., Riezler, S.: QUality estimation from ScraTCH (QUETCH): deep learning for word-level translation quality estimation. In: Proceedings of the Tenth Workshop on Statistical Machine Translation, September 2015
 9. Martins, A.F.T., Astudillo, R., Hokamp, C., Kepler, F.: Unbabel's participation in the WMT16 word-level translation quality estimation shared task. In: Proceedings of the First Conference on Machine Translation: Volume 2, Shared Task Papers (2016)
10. Martins, A., Junczys-Dowmunt, M., Kepler, F., Astudillo, R., Hokamp, C., Grundkiewicz, R.: Pushing the limits of translation quality estimation. Transactions of the Association for Computational Linguistics (2017)
11. Sennrich, R., Haddow, B., Birch, A.: Neural machine translation of rare words with subword units. In: Proceedings of the 54th Annual Meeting of the Association for Computational Linguistics (2015)
12. Shah, K., Bougares, F., Barrault, L., Specia, L.: Shef-lium-nn: sentence level quality estimation with neural network features. In: Proceedings of the First Conference on Machine Translation (2016)
13. Snover, M., Dorr, B., Schwartz, R., Micciulla, L., Makhoul, J.: A study of translation edit rate with targeted human annotation. In: Proceedings of Association for Machine Translation in the Americas (2006)
14. Vaswani, A., et al.: Attention is all you need. Advances in Neural Information Processing Systems 30 (2017)

Tencent Submissions for the CCMT 2020 Quality Estimation Task

Zixuan Wang, Haijiang Wu, Qingsong Ma$^{(\boxtimes)}$, Xinjie Wen, Ruichen Wang, Xiaoli Wang, Yulin Zhang, and Zhipeng Yao

PCG and CSIG, Tencent Inc, Shenzhen, China
{zackiewang,harywu,qingsongma,jasonxjwen,ruichenwang,evexlwang,
elwinzhang,neokevinyao}@tencent.com

Abstract. This paper presents our submissions to CCMT 2020 Quality Estimation (QE) sentence-level task for both Chinese-to-English (ZH-EN) and English-to-Chinese (EN-ZH). We propose new methods based on the predictor-estimator architecture. For the predictor, we propose XLM-predictor and Transformer-predictor. XLM-predictor novelly produces two kinds of contextual token representation, i.e., mask-XLM and non-mask-XLM. For the estimator, both RNN-estimator and Transformer-estimator are conducted and two novel strategies, i.e. top-K strategy and multi-head attention strategy, are proposed to enhance the sentence feature representation. We also propose new effective ensemble technique for sentence-level predictions.

Keywords: Quality Estimation · Predictor-estimator · XLM · Ensemble

1 Introduction

Machine Translation (MT) has achieved great improvement with the development of Deep Learning (DL), which requires accurate and accessible evaluation to further promote the quality of MT outputs. The widely used MT metric BLEU [7] can quickly evaluate the quality of MT outputs, on condition that the human generated reference translation is provided in advance. However, high-quality reference translations demand labor and time. Quality Estimation (QE) becomes an alternative method to evaluate the quality of MT outputs with no access to reference translations [2,11].

Our submissions focus on the sentence-level sub-task of the CCMT 2020 QE Shared Task in both English-to-Chinese (EN-ZH) and Chinese-to-English (ZH-EN) directions. The sentence-level task aims to predict the Human-targeted Translation Edit Rate (HTER) [8] of the MT output, which reflects the minimal amount of edits that is needed to process the MT output to an acceptable level, thus denotes the overall quality of the MT output.

© Springer Nature Singapore Pte Ltd. 2020
J. Li and A. Way (Eds.): CCMT 2020, CCIS 1328, pp. 123–131, 2020.
https://doi.org/10.1007/978-981-33-6162-1_12

Sentence-level QE is commonly formulated as a regression problem. The classical baseline model QuEst++ [9] constructed rule-based features and employed machine learning algorithm to predict HTER scores. Recent neural networks applied the newly-emerged predictor-estimator architecture to QE tasks. Kim et al. [5] proposed the predictor-estimator model first. The predictor aims to extract feature vectors by incorporating large parallel data into a bilingual RNN model, which is subsequently fed into the main bidirectional RNN model to predict QE scores. Later, Fan et al. [1] replaced the RNN-based predictor with a bidirectional Transformer and added 4-dimensional mis-matching features. NiuTrans [10] used Transformer-DLCL based predictor, whereas Unbabel [12] introduced BERT and XLM pretrained predictor models. Besides, ensemble technique emerges as a new trend that can further improve the QE performance. The ensemble approach achieved the best results in the sentence-level QE sub-task of both CCMT 2019 [11] and WMT 2019 [2].

We submit a predictor-estimator based QE system, which extends the open-source OpenKiwi framework[1] [4] to take advantage of recently proposed pre-trained models by transferring learning techniques. Our contributions are as follows:

- We implement two predictors as feature extractors, the Transformer-based predictor (Transformer-predictor) [1], and the XLM-based predictor (XLM-predictor) [6] via the transfer learning technique. For XLM-based predictor, it produces two kinds of contextual token representation in a novel fashion, i.e., masked representations and non-masked representations.
- In addition to the LSTM-based estimator (LSTM-estimator), we use transformer neural networks to build a Transformer-based estimator (Transformer-estimator). We propose novel strategies to optimize the sentence features, i.e., top-K strategy and multi-head attention strategy.
- We ensemble several single-models by regression algorithms to produce a single sentence-level prediction, which outperforms the commonly-used arithmetic average.

2 Architecture

We employ the predictor-estimator architecture built upon the OpenKiwi framework. We adopt XLM-predictor and Transformer-predictor respectively to extract contextual feature vector of the MT output, which could reflect semantic information between the source and the MT output. We innovatively propose mask-XLM and non-mask-XLM, which will be demonstrated in detail below. For the estimator, similarly, different models are used. We adopt LSTM-estimator and Transformer-estimator. Two effective sentence representation strategies for LSTM-estimator are proposed.

---

[1] https://github.com/Unbabel/OpenKiwi.

2.1 Predictors

2.1.1 XLM-Predictor

The Cross-lingual Language Model (XLM) achieved state-of-the-art performances on several Natural Language Preprocessing (NLP) tasks [6]. We extend XLM to QE task and propose novel XLM-predictor.

First, we fine-tune XLM with both Masked Language Modeling (MLM) and Translation Language Modeling (TLM) using large-scale parallel data following the XLM instructions[2].

Instead of using target word representations produced by the fine-tuned XLM as the predictor output as in Kepler et al. [12], we propose non-mask-XLM representation and mask-XLM representation, and adopt further computation to enhance the feature ability. For non-mask-XLM, all words are fed into the XLM to predict each word's representation, enabling the word itself to help predict its representation. For mask-XLM, one target word is masked one time so that the prediction of the masked target word leverages only the surrounding target words and the source context, without any prior information from itself. Let the length of the target sentence be N, the mask-XLM process is repeated N times and thus all target word representations are generated. We further consider two aspects that influence the word representation. One is the weight of each dimension in the word representation. We continue to use the weight during the fully connected layers in XLM. The other is the language embedding, considering that the word representation is closely related to the corresponding language. Formula 1 describes the final word representation produced by XLM-predictor, which is then fed into the estimator as input features to predict HTER scores.

$$Rep_i = R_i \cdot (W_i + Emb_{lang}) \tag{1}$$

where i refers to the i-th word in the target sentence, R_i refers to the original representation of the i-th word, W_i and Emb_{lang} denote the weight of the i-th word and the language embedding of the target sentence respectively. Rep_i is the final representation of the i-th word.

2.1.2 Transformer-Predictor

Transformer-predictor has been proved effective by Fan et al. [1]. Our predictor follows their bidirectional transformer, which contains three modules: self-attention for the source sentence, forward and backward self-attention encoders for the target sentence, and the re-constructor for the target sentence. The semantic features are extracted by bidirectional transformer and human-crafted mismatching features. Our predictor has made one modification: multi-decoding is used during the machine translation module.

To improve the performance, we integrate a XLM-based model, which simply replace the predictor part by XLM. We take the weighted average the two models as the final sentence-level prediction as shown in formula 2. We set α as 0.8 since we emphasize the transformer-based predictor's contribution and incorporate

---

[2] https://github.com/facebookresearch/XLM.

XLM-based predictor only to further enhance the overall performance.

$$Score = \alpha * Score_{Transformer} + \\ (1 - \alpha) * Score_{XLM} \tag{2}$$

2.2   Estimators

Estimator takes features produced by predictor as the input to predict sentence-level scores of the MT output. We implement a multi-layer LSTM-estimator and a Transformer-estimator, both of which adopt novel strategies to optimize the sentence features.

The last state or the mean pooling of hidden states are usually taken as the sentence representation. However, they both have weaknesses: the last state losses certain information of the whole sentence due to the information decay problem, while the mean pooling distributes the same weights to all hidden states. Actually, the contribution of each word to the sentence features varies, which inspires us to take the concept of weight into consideration. We propose two strategies, top-K strategy and multi-head attention strategy, which computing weights from two different perspectives. The two strategies are illustrated in Fig. 1.

2.2.1   Top-K Strategy

Each hidden state is a word representation vector, and each element of the vector represents one feature dimension. From feature dimension perspective, Top-K strategy forms the sentence features by concatenating top-K elements of each feature dimension. The top-K elements refer to the top-K values among all words of the current focus feature dimension. In a result, the sentence feature is a vector with size K * number of feature dimensions.

2.2.2   Multi-head Attention Strategy

Different from top-K strategy, multi-head attention strategy considers the impact of each word on the sentence features via attention mechanism. For each head, we obtain a vector which is a weighted sum of all the word features. By repeating K times, the final sentence feature is a vector with size K * number of feature dimensions. We demonstrate the computation process as formula 3 and 4,

$$a_{k_i} = softmax(h_i * W_k), \tag{3}$$

$$f_{sent} = [\sum_i \alpha_{1_i} * h_i, \dots, \sum_i \alpha_{k_i} * h_i] \tag{4}$$

where a_{k_i} is attention results of each word (h_i), and f_{sent} is the final sentence feature representation.

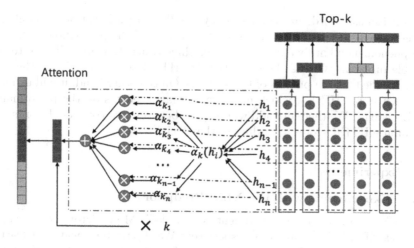

Fig. 1. Sentence representation strategies.

2.3   Ensemble

To boost performance, we ensemble several systems to produce a single sentence score prediction. Model stacking [13, 14] is an efficient ensemble method in which the predictions, generated by using various single systems, are used as inputs of regression algorithm implemented within a two-layer model. To avoid overfitting, we use k-fold cross validation and set k = 5, as described in Martin et al. [15].

We implement and compare several regression algothrims, i.e. Powell's method [16], Quantile Regression, Support Vector Regression (SVR) and Logistic Regression (LR) to optimize for the task metric - Pearson correlation.

3   Experiments and Results

The experiment details below refer to the CCMT 2020 sentence-level QE task only.

3.1   Dataset

The dataset consists of parallel data and QE triplets. Parallel data is used to train the predictor to produce contextual features, which is provided by the CCMT QE task with 8,023,011 EN-ZH parallel sentences (Repeated parallel sentences are filtered). Besides, we use additional 37,128,402 parallel sentences from WMT 2020 task. QE triplets (src, mt, hter) are provided by CCMT QE task, consisting of 10,070 training data (TRAIN) and 1,143 development data (DEV) for ZH-EN, and 14,789 training data and 1,381 DEV for EN-ZH.

We correct one abnormal detail in both the QE TRAIN and DEV triplets for ZH-EN. Take the following sentence as an example: *"Our position is to be courageous, step to be stable. We should not only explore boldly, but also be reliable and prudent, thinking twice before act."*

Two English words are connected by a full stop punctuation without any white-space in the machine translation (MT) file and the post-edited (PE) file. This phenomenon hardly appears in the machine translation and will lead to two possible problems. One is the correctness of HTER scores, which are the gold scores for the training process of QE systems. On the other hand, it will increase Unknown words (UNK), which may exert negative effects on the performance of QE systems. We therefore add white-space between two connected words and re-compute HTER scores according to the official scripts.

3.2 Experiments

3.2.1 Experiments with the XLM-Predictor

For the XLM-predictor, we experiment non-mask-XLM predictor $(non - mask)$ and mask-XLM $(mask)$ predictor respectively. We also try to concatenate feature vectors produced from the two predictors $(Both)$ as the input for the next estimator procedure. Fixing the XLM-predictor, we conduct experiments with LSTM-estimator $(LSTM)$ and Transformer-estimator (TF), each of which adopts multi-head attention strategy $(attn)$ or top-K strategies $(topK)$ to improve the sentence representation.

The results in Table 1 show that our QE systems with XLM predictor achieve moderate correlation with HTER scores in general. On ZH-EN, mask_LSTM_topK ranks top with a Pearson score of .5690, whereas the non-mask_LSTM_attn ranks top with .5329 on EN-ZH. The language features could be an explanation why non-mask-XLM performs better than mask-XLM for Chinese: The Chinese word meaning usually different from that of the consisting characters, so mask one character may affect the word representation.

3.2.2 Experiments with the Transformer-Predictor

We implement a Transformer-based predictor-estimator following Fan et al. [1]. Transformer-predictor has one modification, i.e. the use of multi-decoding during machine translation. To further improve the overall performance, XLM-based predictor is incorporated but with a smaller weight compared to transformer-based predictor as describe in Sect. 2.1.2.

Experiments with the Transformer-Predictor are shown in Table 2, which presents both key configurations and results.

In Table 2, XLM-EST-dim means the dimension in fully connected layer of estimator in XLM-based predictor, while Trans-EST-dim means that in transformer-based predictor. XLM_finetune denotes whether XLM is fine-tuned and XLM-tgt-only means only target information is used in XLM. EST-hidden-dim is the hidden dimension in estimator.

3.2.3 Experiments with Ensemble Methods

We conduct multiple single QE systems through different model architectures or the same architecture with different parameters, and integrate the predictions via stacking ensemble with 4 regressors respectively.

Table 1. Pearson correlations of single QE systems with XLM-Predictor on CCMT 2020 QE EN-ZH and ZH-EN development set for sentence-level task.

Model	ZH-EN	EN-ZH
Both_LSTM_attn	.5468	.5244
Both_LSTM_topK	.5620	.5205
Both_TF_attn	.5364	.4865
Both_TF_topK	.5350	.5056
mask_LSTM_attn	.5542	.4982
mask_LSTM_topK	**.5690**	.4956
mask_TF_attn	.5540	.4951
mask_TF_topK	.5603	.4978
non-mask_LSTM_attn	.5365	**.5329**
non-mask_LSTM_topK	.5507	.5277
non-mask_TF_attn	.5345	.5179
non-mask_TF_topK	.5382	.5208

Table 2. Pearson correlations of single QE systems with Transformer-Predictor on CCMT 2020 QE EN-ZH and ZH-EN development set for sentence-level task.

	Model	Model2	Model3	Model4	Model5
XLM-EST-dim	5140	5140	5140	0	0
Trans-EST-dim	5140	5140	5140	5140	5140
XLM_finetune	1	1	0	1	1
XLM-tgt-only	0	1	1	1	1
EST-hidden-dim	512	256	256	256	512
Pearson-ZH-EN	**.549**	.547	**.549**	.512	.51
Pearson-EN-ZH	.491	**.495**	.491	.456	.453

We select 24 systems based on XLM-predictor and 5 based on Transformer-predictor, then filter single systems with a Pearson score less than 0.5 during ensembling, which leads to 13 systems for EN-ZH, 12 systems for ZH-EN on DEV and 11 system for ZH-EN on PSEU_DEV respectively. 4 regressors refer to Powell's, Quantile Regression, SVR and LR.

Results on DEV with filtered systems are shown in Table 3 prove the effectiveness of ensemble, compared with results shown in Table 1 and Table 2. From Table 3, we also conclude that regression algorithms outperform the simple averaging of single system predictions ("Average" in Table 3).

Table 3. Pearson correlations of ensemble QE systems on CCMT 2020 QE EN-ZH and ZH-EN development set for sentence-level task.

Ensemble methods	ZH-EN	EN-ZH
Average	.5648	.5408
Powell's	.5839	**.5592**
Quantile Regression	**.5848**	.5530
SVR	.5643	.5449
LR	.5843	.5588

4 Conclusion

We describe our submissions to CCMT 2020 QE sentence-level task. Our systems are based on predictor-estimator architecture and built upon OpenKiwi framework. We implement two predictors, Transformer-predictor and XLM-predictor. XLM-predictor novelly produces two kinds of contextual token representation, i.e., masked representations and non-masked representations. Both RNN-estimator and Transformer-estimator are conducted to predict the MT output scores by using the features produced from predictor. Two novel strategies, i.e. top-K strategy and multi-head attention strategy, are proposed to enhance the sentence feature representation. Stacking ensemble is also proved to be more effective than simple averaging integration.

References

1. Fan, K., Wang, J., Li, B., Zhou, F., Chen, B., Si, L.: "Bilingual Expert" Can Find Translation Errors. In: Proceedings of the AAAI Conference on Artificial Intelligence, vol. 33, pp. 6367–6374 (2019)
2. Fonseca, E., Yankovskaya, L., Martins, A.F., Fishel, M., Federmann, C.: Findings of the WMT 2019 shared tasks on quality estimation. In: Proceedings of the Fourth Conference on Machine Translation, vol. 3, pp. 1–10. ACL, Florence (2019)
3. Kepler, F., et al.: Unbabel' s participation in the WMT19 translation quality estimation shared task. In: Proceedings of the Fourth Conference on Machine Translation, pp. 78–84. ACL, Florence (2019)
4. Kepler, F., Trénous, J., Treviso, M., Vera, M., Martins, A.F.: OpenKiwi: an open source framework for quality estimation. In: Proceedings of the 57th Annual Meeting of the Association for Computational Linguistics: System Demonstrations, pp. 117–122. ACL, Florence (2019)
5. Kim, H., Jung, H.Y., Kwon, H., Lee, J.H., Na, S.H.: Predictor-estimator: neural quality estimation based on target word prediction for machine translation. ACM Trans. Asian Low-Resource Lang. Inf. Process. **17**(1), 1–22 (2017)
6. Lample, G., Conneau, A.: Cross-lingual Language Model Pretraining. In: Advances in Neural Information Processing Systems 32, pp. 7059–7069. NeurIPS, Vancouver (2019)

7. Papineni, K., Roukos, S., Ward, T., Zhu, W.J.: BLEU: a method for automatic evaluation of machine translation. In: Proceedings of the 40th Annual Meeting of the Association for Computational Linguistics, pp. 311–318. ACL, Philadelphia (2002)

8. Snover, M., Dorr, B., Schwartz, R., Micciulla, L., Makhoul, J.: A study of translation edit rate with targeted human annotation. In: Proceedings of Association for Machine Translation in the Americas, pp. 223–231. AMTA, Cambridge (2006)

9. Specia, L., Paetzold, G., Scarton, C.: Multi-level translation quality prediction with QuEst++. In: Proceedings of ACL-IJCNLP 2015 System Demonstrations, pp. 115–120. ACL-IJCNLP, Beijing (2015)

10. Wang, Z., et al.: NiuTrans submission for CCMT19 quality estimation task. In: Huang, S., Knight, K. (eds.) CCMT 2019. CCIS, vol. 1104, pp. 82–92. Springer, Singapore (2019). https://doi.org/10.1007/978-981-15-1721-1_9

11. Yang, M., et al.: CCMT 2019 machine translation evaluation report. In: Huang, S., Knight, K. (eds.) CCMT 2019. CCIS, vol. 1104, pp. 105–128. Springer, Singapore (2019). https://doi.org/10.1007/978-981-15-1721-1_11

12. Kepler F, Trénous J, Treviso M, et al.: Unbabel's Participation in the WMT19 Translation Quality Estimation Shared Task. arXiv preprint arXiv:1907.10352 (2019)

13. Wolpert, D.H.: Stacked generalization. Neural Netw. **5**(2), 241–259 (1992)

14. Breiman, L.: Stacked regressions. Mach. Learn. **24**(1), 49–64 (1996)

15. Martins, A.F.T., Junczys-Dowmunt, M., Kepler, F.N., et al.: Pushing the limits of translation quality estimation. Trans. Assoc. Comput. Linguist. **5**, 205–218 (2017)

16. Powell, M.J.D.: An efficient method for finding the minimum of a function of several variables without calculating derivatives. Comput. J. **7**(2), 155–162 (1964)

Neural Machine Translation Based on Back-Translation for Multilingual Translation Evaluation Task

Siyu Lai, Yueting Yang, Jin'an Xu$^{(\boxtimes)}$, Yufeng Chen, and Hui Huang

School of Computer and Information Technology, Beijing Jiaotong University, Beijing, China
{20120374,20120442,jaxu,chenyf,18112023}@bjtu.edu.cn

Abstract. This paper presents the systems developed by Beijing Jiaotong University for the CCMT 2020 multilingual translation evaluation task. For this translation task, we need to build a Japanese-English translation system based on only Japanese-Chinese and English-Chinese data. Our method mainly relies on synthetic data generated by back translation. We implemented three different architectures, namely Transformer-big, Transformer-base and Dynamic-Conv. We also implemented multi-model ensemble technique to further boost the final result. Experiments show that our machine translation system achieved high accuracy without relying on any bilingual training data.

Keywords: Machine translation · Multilingual machine translation

1 Introduction

This paper introduces in detail the submission of Beijing Jiaotong University to the multilingual translation evaluation task in the 16th China Conference on Machine Translation (CCMT2020). This task requires us to build a Japanese-English translation system based on only Japanese-Chinese and English-Chinese data. Due to the lack of direct training data, many techniques wildly used in the area of bilingual translation can not easily be applied in this scenario.

To train a translation model from Japanese to English, we created massive synthetic data based on two MT models of two different directions, namely Chinese-Japanese and Chinese-English [1]. Despite the lack of training data from Japanese to English, the training data for Chinese-Japanese and Chinese-English is readily accessible. The synthetic data can be obtained by translating the Chinese sentences to English of the Chinese-Japanese data, and the Chinese sentences to Japanese of the Chinese-English data. Further cleaning is applied to alleviate the noise contained in our synthetic data.

For the final Japanese-English model, we built our system based on three different architectures, the first one is solely based on attention mechanisms, namely the Transformer model [2]. We further augmented Transformer with deeper encoder layers, to better extract features from source segments, which is named as Transformer-big [3]. Transformer-big was proved to outperform Transformer-base model in most cases.

© Springer Nature Singapore Pte Ltd. 2020
J. Li and A. Way (Eds.): CCMT 2020, CCIS 1328, pp. 132–141, 2020.
https://doi.org/10.1007/978-981-33-6162-1_13

Additionally, we also tried to substitute the self-attention layer with dynamic convolution, providing us another different model to use when doing model ensemble [4].

Then, we applied sub-word segmentation to both languages to resolve the unknown words problem [5], as well as model ensemble, to leverage multi-models to further improve the result [6]. Moreover, we did some contrast experiments, and the results show that our machine translation systems achieved high accuracy without relying on any bilingual data, performed better than other models, proving the effectiveness of our training procedure.

2 Related Work

Recent several years, neural machine translation (NMT) [7, 8] performs end-to-end translation based on an encoder-decoder framework and works well in many machine translation tasks. In this framework, the encoder firstly transforms the source sentence into a fix-length vector, and the decoder generates a target sentence. Such framework has achieved significant improvements over traditional SMT with abundant parallel data available.

However, high-quality and large-scale parallel corpora are non-existent for most circumstances. Lots of work have been done to address this problem, which can be divided into two broad categories: *multilingual* and *pivot-based* approaches.

Johnson [9] has proposed a universal NMT model to translate between multiple languages without any changes in the model architecture, which took full use of multilingual data to improve NMT for all relevant languages. Firat [1] has proposed a multilingual model consisting of several encoders and decoders with finetuning algorithm. It was really difficult to learn and analyze the universal representation for multiple languages, although they have completed direct source-to-target translation without using parallel corpora.

Another crucial way is to bring in a third language named *pivot*, acting as a bridge between the source and the target language. Although it is hard to find available in-domain parallel data, the parallel corpora with a pivot language usually exist. For instance, the parallel data between Japanese and English directly is rare, but the parallel data between Chinese and Japanese, Chinese and English is relatively rich. In pivot-based machine translation, sentences are translated from the source language into pivot language firstly, then translated from the pivot language into the target language.

Although pivot-based method performs well in most translation tasks, it still has some disadvantages. Firstly, the mistakes made in source-to-pivot translation will be propagated to pivot-to-target step which can cause error propagation problem. Furthermore, translated by this pipeline way may lose some relevant information in the pivot translation and may not be represented in the target sentence.

The inspiration for our work came from Sennrich [10], and we introduced a pivot language, via which we could make full use of large bilingual corpora. By back translation, we could create synthetic data for training final model.

3   Model

As we explained above, we combined three different architectures in our work—Transformer-base, Transformer-big and Dynamic-Conv.

3.1   Transformer-Base

The first model we used is Transformer-base, which is a completely attention-based structure for dealing with problems related to sequence models, such as machine translation. The Transformer model dispenses with any CNN or RNN structure, capable of working in the process of highly parallelization, so the training speed is very fast while improving the translation performance.

The structure of Transformer is shown in Fig. 1. The model is divided into two parts: encoder and decoder. The encoder consists of six identical layers, each with two sublayers. The first sub-layer is the self-attention layer, and the second sub-layer is a simple fully connected feedforward network.

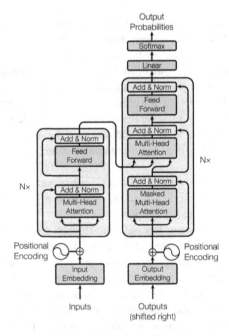

Fig. 1. The Transformer - model architecture.

Residual connections are added outside the two layers, and then layer normalization is performed. In addition to the two layers in the encoder, the decoder also adds a third sub-layer to connect the encoder and decoder. As shown in the figure, the decoder also uses residual error and layer normalization. The output of each sub-layer is:

$$output = \text{LayerNorm}(x + (\text{SubLayer}(x))) \tag{1}$$

The particular attention is called *Scaled Dot-Product Attention*, which takes queries keys of dimension d_k and values of dimension d_v as input, calculates the attention function on a set of queries simultaneously, and packs them together into a matrix Q. The keys and values are also packed together into matrices K and V. The output of the matrix can be calculated as:

$$\text{Attention}(Q, K, V) = \text{softmax}\left(\frac{QK^T}{\sqrt{d_k}}\right)V \tag{2}$$

Multi-head attention allows each head to acquire separate attention weights from different representation subspaces at different position.

$$\text{MultiHead}(Q, K, V) = \text{Concat}(head_1, \ldots, head_h)W^O$$
$$\text{where } head_i = \text{Attention}\left(QW_i^Q, KW_i^K, VW_i^V\right) \tag{3}$$

Where the projections are parameter matrices $W_i^Q \in \mathbb{R}^{d_{model} \times d_k}, W_i^K \in \mathbb{R}^{d_{model} \times d_k}, W_i^V \in \mathbb{R}^{d_{model} \times d_v}$ and $W^O \in \mathbb{R}^{hd_v \times d_{model}}$ (Fig. 2).

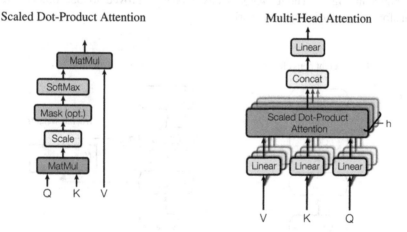

Fig. 2. (left) Scaled Dot-Product Attention. (right) Multi-Head Attention.

Since transformer model does not use any CNN or RNN structure, they introduce some information with relative or absolute position of tokens in the sequence, in order to take advantage of the order information. The position encoding is defined as:

$$PE_{(pos, 2i)} = \sin\left(pos/10000^{2i/d_{model}}\right)$$
$$PE_{(pos, 2i+1)} = \cos\left(pos/10000^{2i/d_{model}}\right) \tag{4}$$

Where *pos* is the position and *i* is the dimension.

3.2 Transformer-Big

To boost its ability to extract features and provide a better presentation for source segment, we deepen the encoder layers for Transformer, and this model is called Transformer-big. Our Transformer-big contains 12 layers of encoder, while Transformer-base only contains 6 layers. However, more encoder layers may encounter the vanishing-gradient problem and entail extra strategies.

3.3 Dynamic-Conv

Self-attention is an effective mechanism. Since it was proposed, it has been applied to many NLP tasks with good performance improvement. However, for long sequences, self-attention is limited by its $O(n^2)$ complexity. In addition, the feature that self-attention can efficiently capture long-term dependence has also been questioned. Therefore, a new structure called *lightweight convolution* is proposed to replace self-attention with CNN structure.

Lightweight convolution uses the prototype of deep (separable) convolution in CV domain, which greatly reduces the number of parameters and complexity by sharing parameters in the channel dimension. Based on the Lightweight, dynamic convolution is proposed, where the weight of CNN is calculated dynamically from the input feature, as shown in Fig. 3. The Dynamic-Conv model is proved to be competitive with Transformer model in many scenarios.

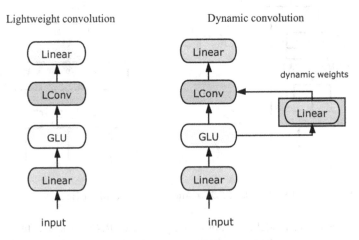

Fig. 3. (left) Lightweight convolution. (right) Dynamic convolution.

$$DynamicConv(X, i, c) = LightConv\left(X, f(X_i)_{h,:}, i, c\right) \qquad (5)$$

Where f is a simple linear module with learned weights $W^Q \in \mathbb{R}^{H \times k \times d}$, i.e.,

$$f(X_i) = \sum_{c=1}^{d} W^Q_{h,j,c} X_{i,c}.$$

4 Experiments

4.1 Preprocessing

Since the quality of training data is vital for the final system, we cleaned the provided training data according to the following strategies:

1. Remove sentences containing too many garbage characters;
2. Remove sentences too long or too short;
3. Remove sentence-pairs with a length ratio too big or too small;
4. Remove duplicate sentence-pairs;
5. Remove sentence-pairs with a too low alignment score provided by fast-align[1];

Both Chinese-Japanese and Chinese-English parallel corpora provided by the organizer contained 3 million sentence pairs. After doing the 5 preprocessing steps mentioned above, 2.99 million sentence pairs were left in each dataset.

And then we used Jieba[2] to perform Chinese word segmentation, and nltk[3] to tokenize English, and Mecab[4] to do Japanese word segmentation. To alleviate the out-of-vocabulary problem and reduce the vocabulary size, we applied sub-word segmentation for both languages, provided by subword-nmt[5].

4.2 Back Translation Based Synthetic Data

To train a translation model from Japanese to English, the parallel corpus from Japanese to English is in need. However, only the parallel data of Chinese-English and Chinese-Japanese are provided. To create synthetic data for the training of final Japanese-English model, we used back translation.

Back translation does not need to make any change in the training algorithm and the model network structure. Back translation has been proved to be simple but effective, while sometimes we may get particularly outrageous translation results in the process of back-translation. Our whole training procedure contains the following steps:

1. Train a Chinese-English model and a Chinese-Japanese model using the parallel data provided.
2. Translate the Chinese sentences in Chinese-Japanese data into English using Chinese-English MT model.

---

[1] https://github.com/clab/fast_align.

[2] https://github.com/fxsjy/jieba.

[3] http://www.nltk.org/.

[4] https://github.com/SamuraiT/mecab-python3.

[5] https://github.com/rsennrich/subword-nmt.

3. Translate the Chinese sentences in Chinese-English data into Japanese using Chinese-Japanese MT model.
4. Combine the synthetic Japanese-English data of step 2 and step 3 together, and train the final Japanese-English model (Fig. 4).

The training steps above were implemented on all of three architectures. Since the synthetic data was generated by our own machine translation model, which means the translated side contained a large amount of noise, we performed the following cleaning

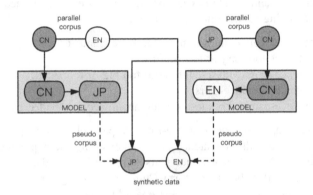

Fig. 4. Back-Translation procedure

steps:

1. Remove sentence-pairs with low language model scores on the target side provided by kenlm[6];
2. Remove sentence-pairs with low alignment scores provided by fast-align;

In the first step, we kept sentences with kenlm scores from -10.0 to -200.0. In the second step, we kept sentences with fast-align scores greater than -500.0. After combining two synthetic datasets, we finally got 5.81 million sentence pairs as training dataset of Japanese-English model.

There are also other ways to deal with the absence of bilingual data, such as pipelined-training and hybrid-labels [9]. Previous evaluation participants and the contrast experiments we did demonstrated that the back-translation based method is normally the most effective while easy to implement.

4.3   Multi-model Ensemble

For multi-model ensemble, we tried the strategy of Independent Parameter Ensemble (IPE), that is to firstly train several models with different architecture and different initialized parameters, and then weight the average probability distribution of multiple

---

[6] https://github.com/kpu/kenlm.

models in the Softmax layer. Better models are assigned with relatively higher weights, and worse models with relatively lower weights.

4.4   Contrast Experiments

In order to prove the superiority of the back-translation based model more comprehensively, we did some contrast experiments like pipeline method, sequence-level knowledge distillation [11] and domain adaptation [12]. Pipeline method firstly translated Japanese to Chinese, and then from Chinese to English. Knowledge distillation used right-to-left and target-to-source model to decode training data, then combining it with synthetic Japanese-English data we generated previously, and used this new data to train model from scratch. Domain adaptation used English sentences in synthetic Japanese-English data to train in-domain model and used English monolingual corpus to train general-domain model, calculating the absolute value of the subtraction between two language models and remaining corpus with low difference value.

4.5   Results

Experiment results on the development set are shown in Table 1. (evaluated by sacreBLEU)

Official automatic evaluation results are shown in Table 2.

4.6   Model Analysis and Discussion

Table 1. Experiments on Development Set

Method	Model	BLEU
Pipeline	Transformer-base	31.29
Knowledge Distillation	Transformer-base	34.47
Domain Adaptation	Transformer-base	34.90
Back-Translation	Transformer-big	35.11
	Transformer-base	35.24
	Dynamic Conv	34.9
	Ensembled	**35.66**

As shown in Table 1, it's obvious that back-translation based model did better than other methods. The reason is that the pipeline method will cause error propagation and

Table 2. Official automatic evaluation results

	BLEU5-SBP	BLEU5	BLEU6
je-2020-bjtu_nlp-primary-a	38.29	40.47	35.81

lose some relevant information, but back-translation based method does not. As for knowledge distillation, because our task is based on patent domain, it is possible that the teacher model is not strong enough to guide the student model. For domain adaptation, remaining corpus with low difference value cannot guarantee the quality of the data. It is possible that both of two language model have low scores and the difference value is relatively small, so the remaining corpus may affect the quality of data. Moreover, using selected monolingual corpus to generate pseudo corpus may damage the quality of data again. Therefore, we can conclude that back-translation is a simple and effective approach to multilingual translation task.

5   Conclusion and Future Work

In this paper, we described our submission in multilingual translation evaluation task. For this translation task, we need to build a Japanese-English translation system based on only Japanese-Chinese and English-Chinese data. Our method mainly relies on synthetic data generated by back translation. We implemented three different architectures, namely Transformer-big, Transformer-base and Dynamic-Conv. We also implemented multi-model ensemble technique to further boost the final result. Experiments show that our machine translation system achieved high accuracy without relying on any bilingual training data.

We have to mention that we also tried other strategies which are commonly used in bilingual translations, including domain adaptation, sequence-level knowledge distillation and checkpoint ensemble [13], but none of them made it to introduce any improvement. Even multi-model ensemble could only introduce minor improvements. This may be caused by the pivot-based synthetic data, and we will explore this problem in our future work.

Acknowledgement. This work is supported by the National Natural Science Foundation of China (Contract 61976015, 61976016, 61876198 and 61370130), and the Beijing Municipal Natural Science Foundation (Contract 4172047), and the International Science and Technology Cooperation Program of the Ministry of Science and Technology (K11F100010).

References

1. Firat, O., Sankaran, B., Al-Onaizan, Y., Vural, F.T.Y., Cho, K.: Zero-resource translation with multi-lingual neural machine translation. arXiv preprint arXiv:1606.04164 (2016)
2. Vaswani, A., et al.: Attention is all you need. In Advances in Neural Information Processing Systems, pp. 5998–6008 (2017)
3. Wang, Q., Li, B., Xiao, T., Zhu, J., Li, C., Wong, D.F., Chao, L.S.: Learning deep transformer models for machine translation. arXiv preprint arXiv:1906.01787 (2019)
4. Wu, F., Fan, A., Baevski, A., Dauphin, Y.N., Auli, M.: Pay less attention with lightweight and dynamic convolutions. arXiv preprint arXiv:1901.10430 (2019)
5. Sennrich, R., Haddow, B., Birch, A.: Neural machine translation of rare words with subword units. In Proceedings of ACL (2016)
6. Rokach, L.: Ensemble-based classifiers. Artif. Intell. Rev. **33**(1–2), 1–39 (2010)

7. Kalchbrenner, N., Blunsom, P.: Recurrent continuous translation models. In: Proceedings of the 2013 Conference on Empirical Methods in Natural Language Processing. In Proceedings of EMNLP, 1700–1709 (2013)

8. Bahdanau, D., Cho, K., Bengio, Y.: Neural machine translation by jointly learning to align and translate. Comput. Sci. 1–15 (2014)

9. Johnson, M., Schuster, M., Le, Q.V., Krikun, M., Wu, Y., Chen, Z.: Google's multilingual neural machine translation system: enabling zero-shot translation. Trans. Assoc. Comput. Linguistics 5, 339–351 (2017)

10. Sennrich, R., Haddow, B., Birch, A.: Improving neural machine translation models with monolingual data. In: Proceedings of ACL (2016)

11. Kim, Y., Rush, A., M.: Sequence-level knowledge distillation. In: Proceedings of the 2016 Conference on Empirical Methods in Natural Language Processing, Austin, Texas. In Association for Computational Linguistics, pp. 1317–1327 (2016)

12. Axelrod, A., He, X., Gao, J.: Domain adaptation via pseudo in-domain data selection. In: EMNLP 2011 - Conference on Empirical Methods in Natural Language Processing, Proceedings of the Conference, pp. 355–362 (2011)

13. Chen, H., Lundberg, S., Lee, S.I.: Checkpoint ensembles: Ensemble methods from a single training process. arXiv preprint arXiv:1710.03282 (2017)

Author Index

Printed in the United States
By Bookmasters